Ulva reticulata, Halymenia durvillei & Sargassum cristaefolium

Nutritional Characteristics & Bioactive Components

MARIBELLE T. HANANI, PH.D.

Ulva reticulata, Halymenia durvillei & Sargassum cristaefolium

Nutritional Characteristics & Bioactive Components

MARIBELLE T. HANANI
ISBN 9 78219 667685
Published by
Yawman Book Publishing House
Davao City, Philippines
+639219512458

ACKNOWLEDGEMENTS

This book would not have been possible without the guidance and help of several individuals, who in one way or another contributed and extended their valuable assistance in the preparation and completion of this study.

To the Department of Science and Technology-PCMARD and PCAARRD for the financial support; the late Dr. Cesario R. Pagdilao, for his unforgettable words that inspired her a lot;

To her MSU-Sulu family, former chancellor Dr. Adjarail B. Hapas and especially the ever visionary chancellor Dr. Nagder J. Abdurahman, the college of fisheries faculties and friends for their spiritual support; understanding and untiring effort.

Last, but not least, her family and the one above all of us, the Omnipresent Allah, for answering her prayers to give strength and plod on despite the struggle and odds; thank you so much, Dear Almighty.

TABLE OF CONTENTS

INTRODUCTION

This book discusses the nutraceutical composition and properties of three selected marine seaweeds *viz Ulva reticulata, Halymenia durvillei* and *Sargassum cristaefolium* from Patikul Higad, Patikul, Sulu and Paayas, Burgos, Ilocos Norte. The proximate and heavy metal compositions of the seaweeds are discussed.

Qualitative phytochemical screenings are applied, and antimicrobial components of the algal extract were tested against the pathogens, *Escherichia coli* and *Staphylococcus aureus* using paper diffusion. Antioxidant properties of seaweed extracts are assessed using diphenyl-p-picrylhydrazyl.

Aside from determining proximate composition, the book also evaluates the heavy metal concentrations in seaweeds. Among the metals analyzed, zinc registered highest concentration (328.74 mg/kg) in *U. reticulata*. Other evaluated seaweed species had lower concentrations of copper (Cu), mercury (Hg), and lead (Pb). The concentration reveals that these seaweeds could be source of the micronutrient - copper.

In addition, except for *U. reticulata* growing, all other seaweeds are safe for human consumption and could be used as raw material for other purposes such as drug or cosmetics preparation.

Lead and Hg are toxic heavy metals and they were also detected in the seaweeds. *Ulva reticulata* were found to contain seemingly high level of Pb. However, no maximum contaminant level values that were specifically set for Pb or Hg in seaweeds., so there was no guideline or standard value that could serve as reference to determine whether the measured concentration is higher or lower than maximum allowable value.

Seaweed appears to be interesting source for medicinal and phytochemical studies. The Presence of phytochemicals in three seaweeds confirmed the presence of terpenoids in both ethanol and methanol extracts. Terpenoids are important as additive in the food industry and as a pharmaceutical agent in biomedicine.

Tannins, saponins and coumarins were present in *U. reticulata* while alkaloids, flavonoids and xantho-proteins were absent in both methods. This difference can be attributed to solubility of the active component in different solvents and methods of extraction were possible source of variation for the chemical composition and bio-activity of the extracts.

Thin layer chromatographic screening showed the presence of saponins, steroids, anthraquinones, phenols and flavonoids in all the seaweeds studied.

Antimicrobial property of seaweeds showed that *S. cristaefolium* has active zone of inhibition on *E. coli* and *S. aureus*.

Dried samples of *U. reticulata* showed best antioxidant activity while *S. cristaefolium* had least antioxidant activity. Among the seaweeds, the green seaweed, *U. reticulata* exhibited the highest free radical scavenging activity while the brown seaweed, *S. cristaefolium* had the lowest. The scavenging activity of the seaweeds on the DPPH radical is dependent on concentration.

In conclusion, the seaweeds screened in this book possess phytochemicals, antioxidants, and antimicrobial potentials, which may be considered for future applications in medicine, cosmetics and feed industry.

Chapter 1
Overview of the Use of Seaweeds

For thousands of years now, our ancestors consumed diets of edible sea vegetables. People roamed the earth, shorelines and harvested what turned out to be the deepest secret to super health and long life. Edible seaweeds are important sources of elements useful for metabolic reactions in humans and animals, such as enzymatic regulation of lipid, carbohydrate, and protein metabolism (Nisizawa et al., 1987).

Seaweeds contain various inorganic and organic substances that can be used for human health, such as polyphenols, carotenoids, tocopherols, terpenes, ascorbic acid, etc alkaloid (Chanda et al., 2010). They are excellent sources of protein, dietary fiber, vitamins (A, B, B_{12}, C, D and E), riboflavin, niacin, pantothenic acid, essential amino acids, folic acid, and essential fatty acids, as well as minerals, such as Ca, P, Na and K. Due to high amounts of vitamins and minerals present in green seaweeds, they are used widely in agriculture, pharmaceutical, biomedical and nutraceutical industries (Cox et al., 2010).

Seaweeds are also used for improving nutrients in animal feed, cosmetics, herbal medicine and fertilizers (Fleurence, 1999; Marinho-Soriano et al., 2006). Moreover, seaweeds also contain potential bioactive compounds which exhibit antibacterial, antiviral and antifungal properties (Marinho-Soriano et al., 2006).

Currently, seaweeds are used worldwide for many different purposes. Human consumption of seaweeds is common in Asian countries, mainly Japan, China, Korea, Vietnam, Indonesia and Taiwan (Dawes, 1988). About 900 species of seaweeds have been recorded in the Philippines, next to China, Japan, France, Korea, USA and Hong Kong (De Silva, 1992). However, the biodiversity of seaweeds is large but difficult to determine, mainly because of the limited biogeographic inventories worldwide (Norton et al., 1996).

Seaweeds are the only sources of phytochemicals, namely: agar-agar, carrageenan, and algin.They are also used as feedstuff for animal consumption (Anantharaman et al., 2009). The term 'nutraceutical' is coined from 'nutrition' and 'pharmaceutical'. It refers to food or food products that provide health and medical benefits in the prevention and/or therapy (Martin et al., 2002). Around 25 seaweed species have been commercially utilized worldwide, and 150 species are consumed as human food (Padua et al., 2004). Seaweeds are treasures of the sea, and species with economic importance are *Beorgesenia, Caulerpa, Chaetomorpha, Codium, Dictyosphaeria, Enteromorpha, Ulva, Hydroclathrus, Sargassum, Turbinaria, Laurencia, Liagora, Rhodymenia, Acanthophora* and *Halymenia* spp.

In the Philippines, utilization of *Sargassum* by coastal populations include; 1) as cover for fishery products to prevent desiccation and maintain their freshness; 2) as food; 3) as fertilizer, insect repellant, flower inducer, and animal feed; and 4) as a therapeutic drink, among others. *Sargassum* biomass is also believed to have a great potential as feedstuff for alternative energy sources, i.e. biofuel (Montaño et al., 2006). *Ulva* is a protein source for sea bream (*Pagrus major*) and *Halymenia* sp. For carrageenan production (Briones et al., 2004).

Sulu is one of the seaweeds suppliers in western Mindanao. Decades ago, seaweeds were known as food delicacies like *Caulerpa, Eucheuma* and *Gracilaria* spp. They are eaten raw, while the industrial and medicinal uses of extracts from seaweeds were known later by inhabitants. The seaweed industry is considered one of the major industries that benefit the majority of coastal fisherfolks. *Eucheuma* and *Kappaphycus* spp. Are the potential seaweeds used for culture and farm in Sulu. Seaweeds like *Sargassum, Ulva* and *Halymenia* spp. Are just flaking in the shoreline for no one knows their benefits.

The coastline of Ilocos Norte, which runs along the China Sea, has rugged topography. This contributes to the diverse part of the country. The seaweed industry in Ilocos Norte is confined to gathering, and commercial culture is not practiced. The biggest market for dried seaweed is in the public market of Laoag City and the Ilocano "balikbayans". It is their favorite take home or "pasalubong" when they return to their residences abroad (Agngarayngay et al., 1983).

Ecologically, the seaweed commodity benefits the surrounding environment, like conserving coastal areas against various environmentally unfriendly catching activities, like poisons and bombs in catching fish. Biologically, seaweed plays an important role as a primary producer of organic materials and oxygen in the coastal environment. Economically, this is a potential commodity to be developed, considering its nutrient content and pharmaceutical uses. It can be used as foodstuffs such as agar, vegetables, and snacks. It also produces algin materials, carrageenan and fluseran used in pharmaceutical, cosmetic, and textile industries, and fish feed.

This study was conducted to analyze the proximate and heavy metal composition, phytochemical components, antioxidant and antimicrobial screening of seaweeds *viz, Ulva reticulata, Sargassum cristaefolium* and *Halymenia durvillei,* collected from Patikul, Sulu and Burgos, Ilocos Norte, Philippines.

This book discusses **the p**roximate and heavy metal composition of selected marine seaweeds, **the p**hytochemical components of the *U. reticulata, S. cristaefolium* and *H. durvillei,*
The antimicrobial property of algal extracts by paper distribution against pathogens, *Escherichia coli* and *Staphylococcus aureus,* and *the a*ntioxidant properties of the seaweed species extract by diphenyl-p- picrylhydrazyl

Chapter 2
Seaweed's Nutritional Contents And Pharmaceutical Uses

This chapter focuses on procedures used to identify seaweed's nutritional contents and pharmaceutical uses. It begins with the taxonomy and morphology of *U. reticulata, H. durvillei and S. cristaefolium*, followed by author's observations on seaweeds.

Seaweeds are marine algae, saltwater-dwelling, category of "plants". Around 9,800 species belong to three major divisions: 1500 green, 1800 brown, and 6500 red species (Guiry, 2000). Seaweeds originated from different evolutionary processes and are classified in different kingdoms such as Plantae (green), Protista (red), and Chromista (brown). Kingdom Chromista has recently been proposed to encompass the "brown algal line" Phaeophyta, Chrysophyta and Pyrrhophyta (Cavalier-Smith, 2007). Systematic classification is mostly based on the composition of pigments involved in photosynthesis; they require space and a hard substrate to attach and grow. Most seaweeds do not have roots to anchor them to soft substrates such as mud or sand (Guillermo, 2008).

Seaweeds play very important ecological roles in many marine communities. They are a food source for marine animals such as sea urchins, fishes, and some food webs' nutritional bases. Seaweeds also provide shelter and home for numerous fishes, invertebrates, birds, and mammals (Waaland, 1997). In addition to "positive" roles, seaweeds also play critical roles in reef degradation, particularly in ecological "phase shifts" where abundant reef-building corals are replaced by abundant fleshy seaweed (Done, 1992 and McCook, 1999). In addition, seaweeds have a broad range of biological activities such as antibacterial, antifungal, antiviral, antineoplastic, antifouling, anti-inflammatory, anti-tumoric, cytotoxic, and antimitotic activities (Michael et al., 2005).

One of the major current human impacts on seaweed communities arises from changes in water quality. Direct human impacts include inputs of pollutants such as nutrients, sediments, and toxic compounds (Schaffelke et al., 2005). Seaweed copes with mechanical stress by having strong holdfast, flexible stipe and blades and bending towards the substrate as waves move over them.

People accept seaweeds either in fresh or dried forms. They are alternative food for fishers during turbulent weather, which deters anglers from catching fish. Almost all *Ilocanos* are accustomed to eating various seaweeds, not only to those living near seashore but also those in the hinterlands (Agngarayngay et al., 2000).

In the Sulu archipelago, seaweed farming became a major livelihood second to fishing, especially coastal dwellers. Seaweed farming improved the living conditions of the populace and helped solve the insurgency problem (Romero, 2002).

In 2003-2010, local seaweed production posted an annual average growth rate of 9%, higher than the 5% in 1997-2002. This rise in production was attributed to the continuous distribution of quality planting materials from many established seaweed nurseries of the Bureau of Fisheries and Aquatic Resources. Seaweed production is presently declining. In 2012, the country produced seaweeds at

1,751,070.64 MT, 4.88% lower than in 2011(Fig. 1; BAS, 2012).

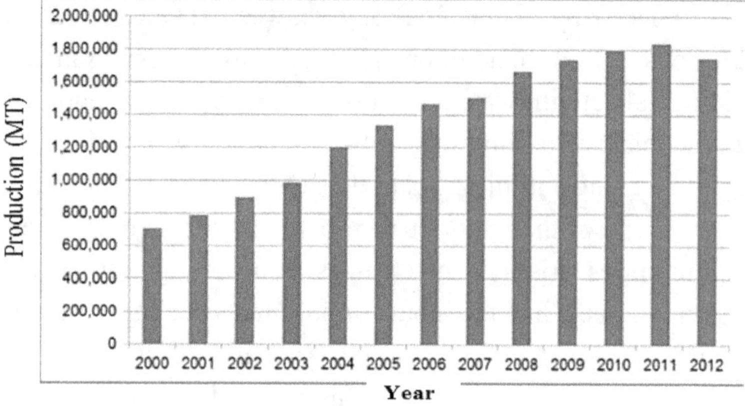

Figure1. Local seaweed production trend (Source: BAS and BFAR).

Taxonomy and Morphology

1. *Ulva reticulata*

Kingdom:	Plantae
Phylum:	Chlorophyta
Class :	Ulvophyceae
Order:	Ulvales
Family:	Ulvaceae
Genus:	*Ulva (naeus)*
Species:	*reticulata orsskal)*
Synonym	*Phycoseris iculata*
Common name	
English	Sea lettuce
Tausug	*Gampal-mpal*
Ilocano	*Gamgamet*

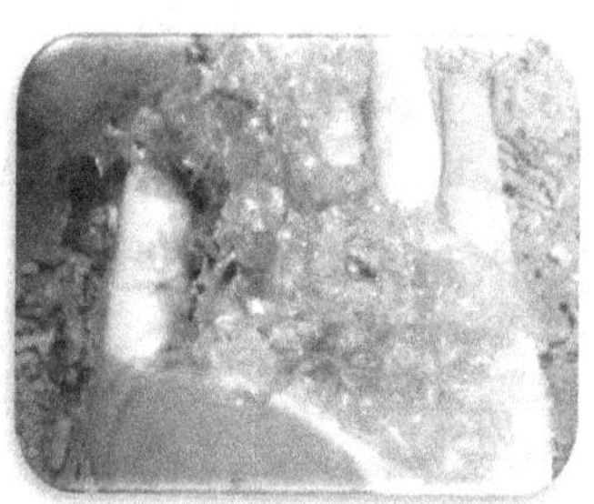

Figure 2. *Ulva reticulata*
(Source: MT Hanani)

Characteristics

Ulva reticulata is a marine edible green alga known as a source of proteins, vitamins, and sulfated polysaccharides. Plants reticulate, have netlike nature obscuring blades, mostly unattached, entangled with other algae. The plant is attached to a substratum throughout its life by holdfast, a disc formed from primary cells of elongated compact and strong nature. The cell contains a single parietal chloroplast, often with a deeply incised or lobed margin containing a single pyrenoid. Chloroplast is mostly located on the outer side of the cell, while the nucleus lies adjacent to the inner wall (Chennubhotla, 1998).

Ulva reticulata is an established tropical species that thrives well in clear, shallow waters. It requires warm water temperature between 25 and 30°C, with growth occurring faster when inorganic nutrients, especially ammonia and phosphorous, are high due to their efficient nutrient uptake ability (Ahmad et al., 2011). *U. reticulata* is conspicuously present during peak growth in protected marine habitats, estuaries, bays, and lagoons, where salinity is around 34-35 ppt. The only exception is the Red Sea, where salinities as high as 36.5 to 39 ppt (average of 38 ppt) seem to be at the end of tolerance for *U. reticulata* (Ateweberhan et al., 2005).

Generally, the geographical distribution of *U. reticulata* is influenced mainly by water temperature. This species can be found at a wide range of temperatures from as low as 20 to 30°C to as high as 38°C in the southern part of the Red Sea (Eritrea) during mid-summer (Ateweberhan et al., 2005). In Sri Lanka (Indian Ocean), surface water temperature around the island is between 26 and 28°C (Coppejans et al., 2009), while that in Southeast Asia ranges between 20 and 29°C. Vietnam, which is connected to continental Asia, has surface water temperature in winter of 20-23°C in its northern region and 25-29°C in the southern part (Nang and Dinh, 1998), while in Thailand, Malaysia, and the Philippines, the temperature is normally about 28-30°C (Trono and Ganzon-Fortes, 1988; Lewmanomont, 1998; Phang, 1998).

In Ryukyu Islands (Okinawa Islands), which a probably are the northern geographical limit of *U. reticulata* in the Pacific ocean, water temperature is influenced by the north branch of the Kuroshio current, resulting in a warmer range of approximately 20°C in winter to 29°C in summer, with an annual average of 25°C typical of tropical waters (Ohno and Largo, 1998).

In Venezuela, *U. reticulata* is found near the Caribbean Sea, where the average surface water temperature is 27°C (with a variation of ± 3°C) and salinity of 34.5 – 36.5 ppt (Smith, 1998). In Chile, where *U. reticulata* has been reported by Etcheverry (1960), the temperature in latitudes 18°S (Arica) to 27°S (Caldera) is around 18-20°C (Alveal, 1998) which is similar to temperature at its northern limits in the Pacific ocean (Okinawa Islands).

Morphology

Green seaweeds are found in fresh and marine habitats. They range from unicellular to multicellular, microscopic to macroscopic forms. Their thalli vary from free filaments to definitely shaped forms. The photosynthetic portion of thalli may be moderate to highly calcified appearing in various forms as fan-shaped segments, feather-like or star-shaped branches with teeth or pinnules, clavate or globose branchlets. (Dhargalkar, 2004).

Pigments

Green seaweeds possess photosynthetic pigments like

chlorophyll a and b, contained in special cell structures known as chromatophores. The cell wall of this group is composed of an outer layer of pectin and an inner layer of cellulose. The photosynthetic product of this group is starch (Dhargalkar, 2004).

Reproduction

Ulva usually multiplies using fragments that are accidentally produced from thallus. Vegetative multiplication by fragmentation also occurs through the proliferation of perennial holdfast. Green algae can produce sexually and asexually by forming flagellate and non-flagellate spores. Alternation of gametophytic and sporophytic generation occurs in this group. Unmated gametes can function as asexual reproductive cells, attaching to substrates and growing into new haploid multicellular gametophytes (Graham et al., 2000).

Physiology

Ulva species, like *U. reticulata,* possess antifouling (Harder et al., 2004), inhibitory (Harder and Qian, 2000), and antimicrobial properties (Vairappan and Suzuki, 2000; Karthikaidevi et al., 2009), which are probably strategies that enable the alga to survive competition (Kolanjinathan and Stella, 2011).

2. *Halymenia durvillei*

Kingdom:	Plantae
Phylum:	Rhodophyta
Class:	Florideophyceae
Order	Halymeniales
Family:	Halymeniaceae
Genus:	*Halymenia*
Species:	*durvillei*
Common name	
English	Flame algae or Dragon breath algae
Tausug	*Buka'tuong*
Ilocano	*Aragan -ilek*

Figure 3. *Halymenia durvillei*
(Source: MT Hanani)

Characteristics

Halymenia durvillei is a beautiful, red slippery, free-floating alga that grows very well under good water quality and medium-high water flow. This adds a splash of color to the reef while exporting nutrients and is easy to cultivate and gently tumble or rest on the bottom. This alga is used for human consumption and is a source of carrageenan. It is attached to rocky substrate by its holdfast in lower intertidal to upper sub-tidal areas that are moderately exposed to wave action. Thalli are large and bushy, red-orange or purple, soft cartilaginous, and slimy when fresh. Stipe is short and supports 2 to 4 main axes 5 to 15 mm wide, branching pinnately-alternately two to four times. Their branches are flattened, their diameter decreases with the increasing degree of branching. Thallus reaches about 35 cm in height (Trono, 2001).

Morphology

Halymenia durvillei is exclusively marine seaweeds that vary in size and shape (Trono, 2001). Numerous short lateral branchlets are formed from the margins of branches. The erect blade with a cuneate base is up to 17 cm long and 1.1 cm wide. Erect blades are sub-dichotomously or trichotomously branched, becoming narrower upwards. Branches are beset with marginal spines (Kawaguchi et al., 2005).

Pigments

Red algae contain chlorophyll a and b, carotene – phycoerythrin pigment. The cell wall of this group is composed of the outer layer of pectin and an inner layer of cellulose. The Photosynthetic product of this group is Floridian starch (Dhargalkar, 2004)

Reproduction

Sexual reproduction is very complicated, involving several structures after the fusion of gametes. Some members of this group exhibit biphasic alternation of generation, in which sexual generation alternates with the asexual generation. In contrast, others are tri-phasic with three generations or somatic phases successively following one another (Dhargalkar, 2004).

3. *Sargassum cristaefolium*

Kingdom   Chromista
Phylum   Ochrophyta
Class   Phaeophyceae
Order   Fucales
Family   Sargassaceae
Genus   *Sargassum*
Species   *cristaefolium*
Synonyms   *Sarassum berberifolium*
Common Name
English   Gulfweed; double-bladed sargassum
Tausug   *Bustak-bustak*
Ilocano   *Aragan*

Figure 4. *Sargassum cristaefolium* (Source: MT Hanani)

Characteristics

Sargassum is called gulfweed and is found to occur only in drifts. Numerous berrylike air sacs keep the branching plant afloat. It is a distinctive and specialized group of marine forms, many of which are not found elsewhere. The main branches of these species are cylindrical, with several spines at the basal portion. Leaves are oblong with a cuneate base, acute apex, and margins serrated. Vesicles are elliptical to fusiform at the top. These plants grow to 1-1.5 m high and form dense communities in the middle to lower intertidal zones along shorelines with moderate to calm water movement (Guiry, 1998).

Morphology

Brown algae are exclusively marine forms. From simple, freely branched filaments to highly differentiated forms, they have different forms. They can be distinguished into blades, stipes, and holdfast. *S. cristaefolium* are epiphytes that grow as a crust on rocks or shells as large fleshy, branched or blade-like thalli (Trono, 1992).

Pigments

Photosynthetic pigments of the brown algae are chlorophyll a and c, carotene, xanthophylls, and fucoxanthin. The cell wall is composed of the outer layer of algin and the inner layer of cellulose. Photosynthetic products of the brown algae are laminarin and mannitol (Dhargalkar, 2004).

Reproduction

Brown algae reproduce sexually and asexually. A species reproduces vegetatively by fragmentation and produces biflagellate neutral spores found within one-celled or many-celled reproductive organs. Sexual reproduction is through the union of flagellated male and female gametes. Alternations of gametophytic and sporophytic generations occur in this group except in members of *Fucales* (Dhargalkar, 2004).

CHAPTER 3
NUTRITIONAL PROPERTIES OF SEAWEEDS

Nutritional studies on seaweeds indicate that brown and red seaweeds possess good nutritional quality. Seaweeds can play a vital role in human health and nutrition. Peptides derived from algal protein are most promising as antihypertensive agents. Typical nutritional analyses of seaweeds have identified high levels of carbohydrates as well as minerals, vitamins, and trace elements like iodine (Guiry et al., 2001)

Aside from human food, seaweed is also considered feed for livestock, poultry, and a source of fertilizers (Mohanty et al., 2013). Carbohydrates are principal constituents of seaweeds and contain fats, protein, minerals, and vitamins (Dagoon, 2005).

Ulva reticulata, a marine edible green alga, is a known source of proteins, vitamins, and sulfated polysaccharides (Rao et al., 2004). *Sargassum* extract was tested as fertilizer for chrysanthemum, petchay, strawberries, carrots, potatoes, tomatoes, and a postharvest treatment for tomato and chrysanthemum (Montano and Corpuz, 2002).

Halymenia sp. is a source of carrageenan, agars, and complex sulfated galactan. Agar is a complex polysaccharide made of a neutral fraction called agarose (Philips et al., 2000). Carrageenan is a generic term for a complex family of anionic polysaccharides. Polysaccharide is known for its gelling and thickening properties and is used widely in food, cosmetics, and pharmaceutical properties (Lapasin et al., 1995 and Chopin, 2007).

Proximate Analysis on Nutritional Composition of Seaweeds

Proximate analysis is a chemical method of assessing and expressing a feed's nutritional value, which reports moisture, ash, crude fiber, crude fat, and crude protein present in the fuel as a percentage of dry fuel weight. Carbohydrate is determined by difference. Proximate analyses give the overall nutritional composition of the sample in question; this is briefly complemented by the anti-nutrient and mineral composition of the sample.

Moisture content

Moisture content indicates the presence of moisture in considerable amount and too much moisture in any food sample can make the sample viable for microorganism growth. This accounts for most of the biochemical and physiological reactions in the plant (Guisseppe and Baratta, 2000)

Fiber

The nature of soluble seaweed fibers is that their passage through the gastrointestinal tract occurs largely without digestion. Therefore, fibers can increase feelings of satiety and aid digestive transit through their bulking capacity (Pearson et al., 2005).

Mineral

Seaweeds are high in minerals due to their marine habitat, and the diversity of minerals they absorb is wide. Important minerals, like calcium, accumulate in seaweeds at much higher levels than in terrestrial foodstuffs. Minerals like iron and copper are present in seaweeds at higher levels than in many well-known terrestrial sources of minerals like meats and spinach. For example, there is more iron in an 8g serving of dry *Palmaria palmate* (Dulse Dillisk) than in 100g of raw sirloin steak (6.4 mg versus 1.6 mg, respectively (Mc Cance et al., 1993)

Vitamin

Seaweeds vary from species to species, but many spend large amounts of time exposed to direct sunlight in an aqueous environment. As a result, seaweeds contain many antioxidants, including vitamins and protective pigments. Seaweeds are also one of a few sources of vitamin B_{12}. *Ulva lactuca* can provide this vitamin over recommended dietary allowances for Ireland of 1/4g day with 5g in 8g of dry stuff (Food Safety Authority of Ireland, 1999).

Fatty acid

Seaweeds contain many essential fatty acids, which may add to their efficacy as a dietary supplement or as part of a balanced diet (Dembitsky et al., 2003). Seaweeds contain up to 2% of lipids' dry weight, and much of this lipid content is made up of polyunsaturated fatty acids (PUFA)(Tobin et al., 2007). PUFAs account for almost half of this lipid content, with much of it occurring in the forms of omega-3 and omega-6 lipids. (British Nutrition Foundation, 2007). Seaweeds are also normally tested after drying, but the effects of other types of food processing, like canning, have been found to have a detrimental post-processing effect on fatty acid levels (Sanchez et al., 2004).

Protein

Protein content can be as high as 47% of the dry weight (Fleurence, 1999). Some seaweeds, like *Porphyra* spp. (Nori), are high in proteins. Levels vary according to season and species. For most species, aspartic and glutamic acids constitute a large part of the amino acid make-up of these proteins. Essential amino acids such as histidine, leucine, isoleucine, and valine are present in many types of seaweeds, like *P. palmata* and *Ulva* spp. (sea lettuce). Levels of isoleucine and threonine in *P. palmata* are similar to levels found in legumes and histidine found in *Ulva pertusa* at levels similar to those in egg proteins (Fleurence, 1999). Overall, seaweeds reviews have found them as sources of nutrients and proteins for nutritional purposes (Wong, 2000).

Heavy Metals

Excessive concentrations of trace elements [cadmium (Cd), cobalt (Co), chromium (Cr), mercury (Hg), manganese (Mn), nickel (Ni), lead (Pb) and Zinc (Zn)] are toxic and lead to growth inhibition, decrease in biomass and death of the plant (Zenk, 1996). The main component responsible for metal sorption is alginate which is present in a gel form in algal cell walls (Fourest and Volesky, 1997).

Types and concentrations of metals found in seaweed vary with species, collection time, growth phase, and collection site (Zhou et al., 1998). However, *sargassum* spp. has been identified biosorbent material in an attempt to confirm its high metal-sorbing performance as a property of the seaweed genus (Volesky and Kuyucak, 1988; Volesky and Schiewer, 1999).

High concentration of heavy metals in the human body can cause diseases and illnesses (Eisler, 2000). Recent studies have focused on marine algae, otherwise known as seaweeds (Figueira et al., 2000). Seaweeds are excellent agents of filtering the metals such as Zn, Cd, copper (Cu), Ni, and iron (Fe) (Sanchez et al., 2001).

World Health Organization (WHO) estimated the tolerable concentration of 0.2 µg/m or 200 ppm) on long-term inhalation exposure to elemental mercury vapor and tolerable intake of total mercury of 2 µg/kg (0.002 ppm) body weight per day. Metals considered nonessential (Pb, Cd, Cr, and Hg) are potentially highly toxic for plants (Sebastiani et al., 2004).

Heavy metals are non-biodegradable, have long biological half-lives, and have the potential for accumulation in different body organs leading to acute and chronic toxic effects (Radwan and Salama, 2006). Iron, copper, and magnesium are the most significant metals found in red and green algae (Riosmena et al., 2010).

Phytochemicals

Phytochemicals are non-nutritive bioactive plant substances. Seaweeds are a source of biologically active phytochemicals, including carotenoids, phycobilins, fatty acids, polysaccharides, vitamins, sterols, tocopherol, and phycocyanins, among others. Many of these compounds are known to possess biological activity, hence, have potential beneficial use in health care (Kadam et al., 1980)

Total extraction of material carried out by any polar solvents such as acetone, aqueous methanol (80%), and aqueous ethanol, re-extraction with hexane, chloroform, and ethyl acetate lead to successive extraction of terpenoids and sterols (Harborne, 1998; Bhat, 2005).

Natural products can be mainly divided into three groups such as primary metabolites, secondary metabolites, and high molecular weight polymeric materials (Hanson, 2003). Primary metabolites include nucleic acids, amino acids, and sugars in all cells and play a central role in the metabolism and reproduction of the cells. High molecular weight polymeric materials such as cellulose, lignins, and proteins participate in cellular structure. Natural product term refers to any naturally occurring compounds but in most cases mean secondary metabolite (Hanson 2003; Sarker et al., 2005).

Phytochemical-rich foods should clearly form part of a healthy balanced diet. However, the human body has a number of physiological, biochemical, and enzymatic processes by which it can combat oxidative stress outside of dietary intake. The route by which a wide variety of phenolic compounds enters circulation is not well characterized, nor is the bioavailability and distribution of such factors in the human body. However, previous intervention studies have not shown evidence of parallel change in the body's total antioxidant capacity when the dietary antioxidant intake has increased (Kulling, 2008).

While the above finding casts doubt on the benefit of increasing polyphenolic consumption from the perspective of reducing oxidative stress, it must be noted that such compounds may have other physiological effects. Previous studies in animal models and cell culture have suggested that seaweed phytochemicals can inhibit the progression of carcinoma formation (Apostolidis, 2010).

Qualitative Phytochemical Analysis

Qualitative phytochemical analysis can be performed through preliminary analysis through test tube methods for the following: alkaloids, coumarins, flavonoids, saponins, tannins, terpenoids, and xanthoprotein (Evans et al., 2002).

Alkaloids

Alkaloid is a class of organic compounds made up of carbon, hydrogen, nitrogen, and usually oxygen characterized as "alkali-like" compounds found in marine organisms and marine seaweed (Bently 1957). Alkaloids have a wide range of pharmacological activities, including anti-malarial, antiasthma, and anticancer Kittakoop et al., 2014); cholinomimetic (Russo et al., 2013); vasodilatory, antiarrhythmic, and analgesic (Raymond et al., 2010); and antibacterial (Cushnie et al., 2014).

Anthraquinones

Anthraquinones is a class of natural compounds that consists of several hundreds of compounds that differ in nature and positions of substituent groups (Schripsema et al., 1999). Anthraquinones compound is used as laxative mainly from their glycosidic derivatives and used to treat fungal skin diseases (Li et al., 2004). Derivatives are frequently found in slimming agents and have been valued for their cathartic and presumed detoxifying action; however, they cause nausea, vomiting, abdominal cramps, and diarrhea, with both therapeutic dose and overdose (Li et al., 2004).

Coumarins

Coumarin is a phytochemical with a vanilla-like flavor. Coumarin increases blood flow in veins, has anti-fungicidal and antitumor activities, and decreases capillary permeability. It is used to treat asthma (Liu, 2011) and lymphedema. (Farinola et al., 2005).

The German Federal Institute for Risk Assessment has established tolerable daily intake (TDI) of 0.1 mg coumarin per kg body weight but also advises that higher intake for a short time is not dangerous. The Occupational Safety and Health Administration (OSHA) of the United States do not classify coumarin as a carcinogen for humans.

Flavonoid

Flavonoid has antioxidant activity with very popular health-promoting effects. Activities attributed to flavonoids include anti-allergies, anticancer, antioxidant, anti-inflammatory, and antiviral. Flavonoids quercetin is known for relieving hay fever, eczema, sinusitis, and asthma (Harborne et al., 2000).

Saponins

Saponins have antitumor and anti-mutagenic activities and can lower the risk of human cancers by preventing cancer cells limiting their growth and viability. When ingested by humans, it seems to help the immune system and protects against viruses and bacteria (George et al., 2002).

Steroids

The steroid has an organic compound that contains a characteristic arrangement of four cycloalkane rings joined to each other. Steroids are drugs and help cure illnesses but in allowable and recommended dosage. Steroids are necessary for life at all levels and are needed to treat high cholesterol levels and hormonal imbalance in sex (Lednicer, 2011). They also include drugs like the anti-inflammatory agent dexamethasone are steroids (Rhen et al., 2005).

Tannins

Tannins are anti-nutritional compounds that have antimicrobial activity (Schalbert, 1991). They belong to a group of chemical compounds produced by some broadleaf forage plants that can bind proteins to prevent cardiovascular disease. The risk of atherosclerosis is increased by high blood pressure, hypertension, kidney disorders, obesity, diabetes, smoking, excessive alcohol consumption, stress, thyroid and adrenal gland problems, and lipid disorders (Crespy and Williamson, 2004).

Terpenoids

Terpenoids are generally alicyclic lipid-soluble compounds, and isomerism is common. Due to the twisted cyclohexane ring in chair form, different geometric conformations are possible depending on the substitution around the ring. TLC also allows the isolation of various classes of terpenoids on silica gel and silver nitrate impregnated silica gel coated plates (Harborne, 1998; Bhat, 2005).

These structural features may cause artifact formation during isolation (Harborne, 1998). For example, silver nitrate impregnated silica gel also provides separation of terpenoids containing unsaturation (Bhat, 2005; Sarker et al., 2006). In addition, it has anti-inflammatory and antioxidant activities (Houghton et al., 2003).

Xanthoprotein

Xanthoprotein is a qualitative test for the presence of protein. It is a yellow acid substance formed by hot nitric acid on albuminous or protein matter and is changed to a deep orange-yellow color by adding ammonia (George, 2012).

Phenolic

Phenolic is an aromatic ring bearing one or more hydroxyl substituents and ranges from simple phenolic molecules to highly polymerized compounds (Bravo, 1998). It is ubiquitous in plants which are essential parts of the human diet and is of considerable interest due to its antioxidant properties. Marine seaweeds are rich sources of various phenolic antioxidant compounds (Smit et al., 2004; Randhir et al., 2004).

These compounds exhibit a wide range of physiological properties, such as anti-allergenic, anti-atherogenic, anti-inflammatory, antimicrobial, anti-thrombotic, cardio-protective, vasodilatory, and have antioxidant effects (Athukorala et al., 2006).

These are widely used in manufacturing resins, plastics, insecticides, explosives, dyes, and detergents and as raw materials for producing medicinal drugs like aspirin (Michael, 2008).

Bioactive Components of Seaweed

Seaweeds and seaweed-derived products are underexploited marine bio-resources and natural ingredients for functional foods. Recently, aquatic habitats have increasingly been shown to provide a rich source of natural bioactive compounds with hypo-cholesterolemia, anti-inflammatory, antiviral, antineoplastic, antimicrobial, and hypertensive properties (Hanaa et al., 2008).

Seaweeds have furones that work as a powerful antibiotic by preventing communication. They are effective against a wide range of bacteria with a communication system, like bacteria that cause golden staph infections and tuberculosis. Here are promising briny plants from seaweed to sample, *Nori* (*Porphyra* species), *Kombu* (*Laminaria japonica*), *Arame* (*Eisenia bicyclis*), and *Wakame* (*Undaria pinnatifida*). *Wakame's* pigment, fucoxanthin, is known to improve insulin resistance.

Fucoxanthin burns fatty tissue. *Nori* contains Vitamins C (a potent antioxidant) and B_{12} (crucial for cognitive function), and the compound taurine, which helps control cholesterol. Kelp is also rich in fucoidan, a phytochemical that acts as an anticoagulant; Arame also has antiviral properties with anti-obesity effects. The discovery of new analytical methods and techniques is important for studying metabolites in seaweed concerning their applications in pharmacology and the food industry (Onofrejova et al., 2010).

CHAPTER 4
FOOD VALUE OF SELECTED SEAWEEDS

Proximate analysis in food science estimates the basic composition of food samples in terms of moisture content, crude protein, crude fat, ash or mineral content, total carbohydrates, and total soluble solids. Figure 11 presents the moisture content (%) of air-dried samples of seaweeds collected from Patikul, Sulu, and Burgos, Ilocos Norte. The moisture content ranged from 14.00% ± 0.14 to 24.06% ± 0.25. The highest and lowest moisture contents were noted in *Sargassum cristaefolium* and those from Patikul; Sulu has the highest moisture content (24.06% ± 0.25) while those from Burgos, Ilocos Norte has the lowest value (14.00% ± 0.14).

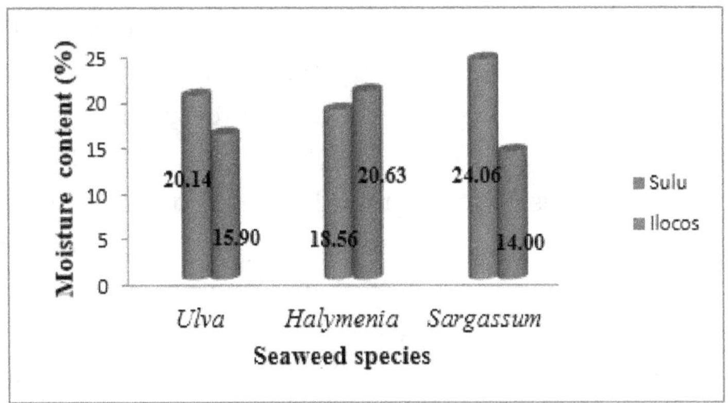

Figure11. Moisture contents of the seaweeds

In terms of location, seaweeds from Patikul, Sulu, particularly *S. cristaefolium* and *U. reticulata* have higher moisture content compared to those from Burgos, Ilocos

Norte. Specifically, the study showed that geographical location, water current, and time of collection vary the moisture content of certain seaweeds. It has been reported in numerous studies also that the biochemical composition of seaweeds varies with species, geographical location, season, and temperature (Khairy and El-Shafay, 2013; Munier et al., 2013).

The moisture content of the food sample reflects the amount of dry matter or solid material that it can take. Higher moisture content means lower solid material or dry matter that it can supply as food and vice versa. This shows that seaweeds with lower moisture content can supply more solid food.

Hence, *S. cristaefolium* collected from Patikul, Sulu, is a good source of dry matter among the three classes of seaweeds because the area is not disturbed by many factors associated with the ecosystems. Seaweed farming in the Sulu archipelago became a major livelihood second to fishing, especially among coastal dwellers. Nevertheless, all the seaweeds are good sources of dry matter.

The protein content of the seaweeds ranged from 3.96% ± 0.03 (*H. durvillei* – Ilocos Norte) to 6.61% ± 0.01 (*U. reticulata* – Ilocos Norte; Figure 12). Among the three classes of seaweeds, green *U. reticulata* had highest protein content with a mean of 5.91% ±0.70 from two locations, followed by the brown *S. cristaefolium* with 4.25% ±0.07 and the lowest from red *H. durvillei* with 4.09% ±0.13. Pise and Sabale (2010), who conducted studies on different seaweeds classes, reported similar results.

In addition, Parthiban et al. (2012) found green seaweeds to have the highest protein content, while the lowest was that of red seaweeds. Regarding geographical location, U. reticulata from Sulu provides the highest protein (6.61%

±0.01) among the seaweeds evaluated. Generally, seaweeds can only provide minimal protein when taken as food. According to Manivannan et al. (2008), the protein contents of Sargassum, Turbinaria and Gracilaria species were less than 10%. In the present study, brown seaweed, *S. cristaefolium* had a protein content of 4.17% from Burgos, Ilocos, Norte, and 4.32% from Sulu, which were higher

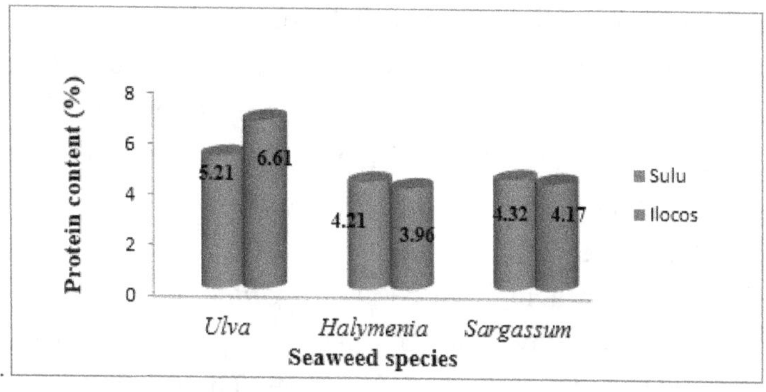

Figure 12. Protein content (%) of the seaweeds

compared to 3.84% ±0.02 from *S. ilicifolium* reported from other studies

The protein content of the seaweeds varies with species and geographical location. David and Parekh (1975) reported significant variations in protein content of the same species of brown seaweeds grown in different locations and periods.

U. reticulata from Ilocos Norte had the highest protein content of 6.61%, maybe because this was collected along the shoreline with a salinity of 35 ppt and had a pH of 9.0. Areas near the shoreline are known to possess nutrients ideal for plant growth. On the other hand, seaweeds collected from Sulu were collected from the open sea in which nutrient content is expected to be lower. The nutrient content is usually higher near shoreline than in open water in the marine environment. Water current is also known to affect growth and the phytochemistry of seaweeds, but water current was not monitored in the two sites.

Seaweeds, as primary producers, enhance organic production and serve as food to various herbivores like siganids and sea urchins, which are likewise food of some animals in a higher level of the food chain; hence, the productivity of reef flat improves (Trono 1993).

The Carbohydrate content of seaweeds ranged from 36.53 to 56.88% (Figure 13). *S. cristaefolium* from Ilocos Norte has the highest carbohydrate content (56.88%) while the lowest was noted in *H. durvillei* also from Ilocos Norte (36.53%). Taking mean in two locations, the highest carbohydrate content is in *S. cristaefolium* with a mean of 47.93% ±8.96 followed by *H. durvillei* at 44.69% ±8.16 and lowest in *U. reticulata*, 42.75% ±1.79. These results of the present study are in contrast with Parthiban et al. (2012), where green seaweed, Enteromorpha compressa, was found to have the highest carbohydrate contents while lowest was in brown seaweed *Dictyota dichotoma*.

In terms of geographical location, carbohydrate content in *U. reticulata* and *H. durvillei* was higher in Sulu than in Ilocos Norte. In *S. cristaefolium, the* carbohydrate content of samples from Ilocos Norte was higher than that from Sulu. However, *U. reticulata* from both locations has an almost similar composition, unlike in *S. cristaefolium* and *H. durvillei,* which have great differences in carbohydrate content.

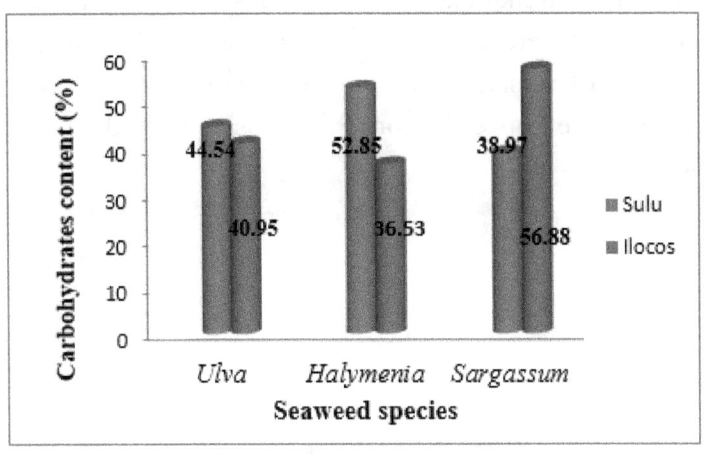

Figure13. Carbohydrate contents of the seaweeds

Results revealed variations in carbohydrate content in the same species in different geographical locations. Furthermore, there are also differences in carbohydrate content among classes of seaweeds (Parthiban et al., 2012; Khairy and El-Shafay, 2013; Manivannan et al., 2008). However, these seaweeds can provide more than 35 to 55% of carbohydrates on a dry basis when taken as food. In addition, carbohydrates are an important source of energy for respiration and other metabolic processes.

Figure 14 presents the lipids content of three seaweeds ranging from 0.32 ± 0.00 to 1.29% ± 0.41. Highest and lowest contents were noted in *S. cristaefolium* but from different geographical location. *S. cristaefolium* in Sulu has 1.29% lipid while that in Ilocos Norte, has 0.32%. Among classes of seaweeds, brown, *S. cristaefolium*, has highest mean lipid content of 0.81% ±0.49 followed by green, *U. reticulata* with 0.56% ±0.12 and lastly the red, *H. durvillei* with 0.41% ±0.00. Lipid content of seaweeds in these locations is very low compared to similar classes of seaweeds studied by Parthiban et al. (2012) where reported values ranged from 3.15 to 5.30%. Khairy and El-Shafay (2013) also noted the lipid content of representative seaweeds from these classes ranged from 1.47 to 4.09%.

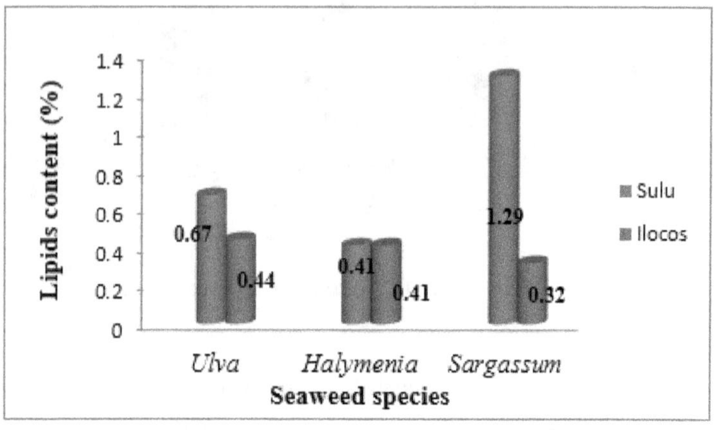

Figure14. Lipid contents of the seaweeds

According to Ratanarpon and Chirapat (2006), seaweeds are not good sources of lipids and they usually contain less than 4% total lipids (Herbeteau et al., 1997). Findings of this study are supported by findings of Khairy and El-Shafay (2013) in *Ulva lactuca, Janiarubens* and *Pterocladia capillacea* (1.47 – 4.09%), Reeta (1993) in *Sargassum wightii* (0.16–1.55%), and Shanmugam and Palpandi (2008) in *U. reticulata* (1.70%). Differences in lipid content can be attributed to geographical location, environmental conditions, salinity, or other water parameters.

Ash content of seaweeds is shown in Figure 15. The ash value in seaweeds ranged from 23.97% ± 0.8 to 38.47% ± 1.12. Values show that these seaweeds are good sources of minerals. The highest and lowest ash value can be noted in *H.*

durvillei from different geographical locations. *H. durvillei* from Ilocos Norte has higher ash content than that from Sulu.

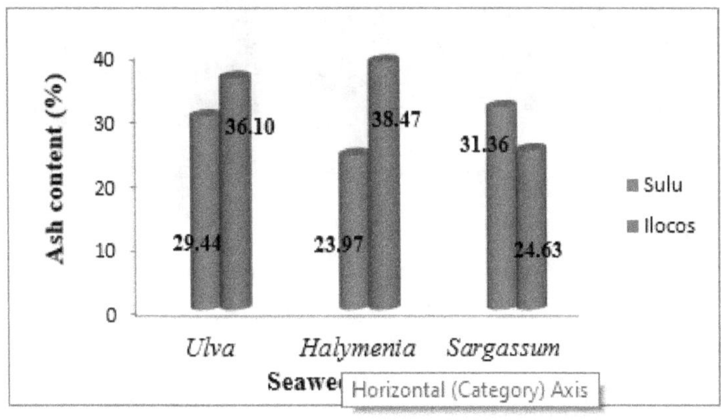

Figure15. Ash contents of the seaweeds

Among classes of seaweeds, red, *H. durvillei* (38.47% ± 1.12) has the highest average ash content in the two locations, followed by green, *U. reiculata* (36.10% ±19.38) and lowest in brown, *S. cristaefolium* (24.63 ± 0.54). These study results are supported by the study of Manivannan et al. (2008), that green seaweed species have higher ash content than brown species. Results show that ash content varies with species and geographical location.

Figure 16 presents the total soluble solids of dried seaweeds. Values obtained ranged from 0.13% ± 0.01(*U. reticulata* - Ilocos Norte) to 0.58% ±0.31 (*S. cristaefolium* - Sulu). Among classes of seaweeds, *S. cristaefolium* was found to have highest amount of total soluble solids averaging 0.58% ±0.31 among seaweeds, followed by *H. durvillei* with 0.40% ±0.01 and least was found in *U. reticulata* with 0.15% ±0.02.

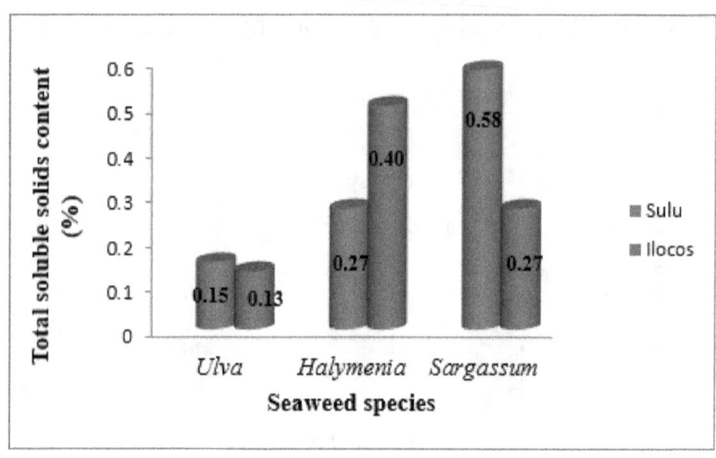

Figure16. Total soluble solids (%) of the seaweeds

S. cristaefolium has the highest total soluble solids among seaweeds collected from Sulu. Conversely, U. reticulata has the lowest total soluble solids among seaweeds collected from Ilocos Norte. Results revealed variations in total soluble solids present within species and geographical locations. The amount of soluble materials in these seaweeds is very minimal, indicating they contain more insoluble fractions and are good dietary fibers that can aid in the digestion process.

Figure 17 shows the amount of the mineral sodium present in the seaweeds. Sodium content of the seaweeds ranged from 7.70% (*H. durvillei* – Ilocos Norte) to 12.50% (*H. durvillei* – Ilocos Norte). These seaweeds can play an important role in the electrolyte balance in humans, especially in terms of sodium content.

The order of sodium content from highest to lowest according to classes based on the mean of the two sites is *H.*

durvillei (10.10% ±2.40), *S. cristaefolium* (1.04% ±0.20), and *U. reticulata* (0.82% ±0.23). However, the sodium content in *H. durvillei* – Sulu and Ilocos Norte are very high for the daily requirement of sodium by humans.

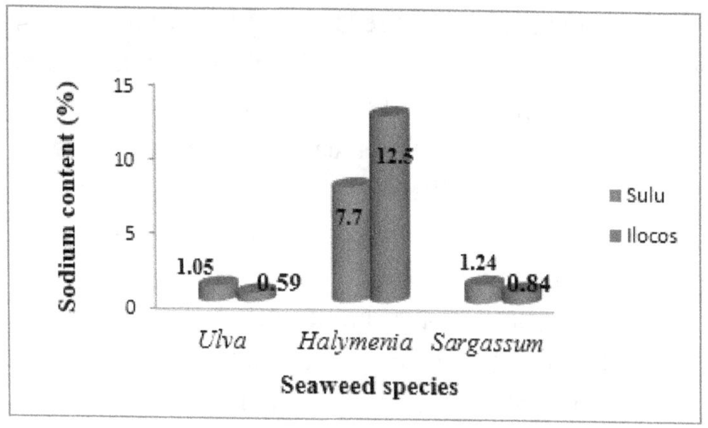

Figure17. Sodium content (%) of the seaweeds

Study of El-Said and El-Sikaily (2013) found red and brown seaweeds to have higher averages of Na and K than green. Similarly, this study found red seaweed to have the highest sodium content, followed by brown seaweed and green seaweed. Results show the ability of different seaweeds to take in different amounts of minerals from water depending on the environment where they are located.

This present study revealed that topography and geographical location affect the phytochemistry of certain seaweed. As we all know, in Burgos, Ilocos Norte, seaweeds are their delicacy, and in most of the seaside settler's collection of seaweeds is their mode of living. Therefore, the constant harvest of seaweeds may deplete their nutrients, thus affecting the long-term viability of seaweed itself.

November of 2013 was the month seaweed collection in Burgos, Ilocos Norte. Weather conditions had affected nutrient composition due to strong current, thereby disturbing the substrate. Due to strong currents, wastes, especially organic ones, accumulate in reefs affecting their nutrient composition, especially the sodium content of seaweed is very high. In comparison, seaweeds in Patikul, Sulu have more appealing qualities as a commodity in appearance, texture, and nutrient composition. These qualities of seaweeds may be attributed to the area of a location not disturbed by any ecosystem factor-like illegal fishing activities, examples of which are dynamite and cyanide fishing which are not common practices in the area.

Both locations have salinity ranges from 34 to 35 ppt and pH of 8 to 9 but differ on collection time and weather conditions. Therefore, this study cannot compare the nutrient composition of the seaweeds based on the result due to some factors stated earlier.

Seasonal variations in chemical composition and nutritional value have been reported in common marine seaweeds. Proteins, carbohydrates, and lipids are algae's most important biochemical components. Lipids are widely distributed in several resistance stages of seaweed. Seaweeds belonging to Rhodophyta possess high proteins (10–30% DW) (Darcy-Vrillon 1993). The protein contents of some red seaweeds, like *Palmaria* and *Porphyra tenera*, are 35 and 47% DW, respectively (Morgan et al., 1980). These levels are comparable to soybeans (35% DW). Overall, seaweeds have been reviewed favorably as sources of nutrients and proteins for nutritional purposes (Wong et al., 2000).

CHAPTER 5
PROXIMATE COMPOSITION OF THE SEAWEEDS AS FEEDS

Ulva reticulata has high protein (17.91%) and ash content (37.38%). *Halymenia durvillei* is high in moisture content (11.07%), while *S. cristaefolium* is high in fiber (8.59%) and ash content (15.07%) on a dry basis (see Figures 18 and 19). Based o the FAO report (2009), the protein contents of the green and red seaweeds are quite variable. Protein contents of the green seaweeds ranged from 6.00 to 26.00%, while red ranged from 3.00 to 29.00%. This study found the protein content of green and red seaweeds to be between these ranges.

Fresh *U. reticulata* from Tanzania was analyzed to have 25.70% crude protein, 18.30% ash, and 38.50% crude fiber (Msuya and Neori, 2002). Oven-dried meal red seaweeds from the Philippines, *Kappaphycus alvarezii,* and *Gracilaria heteroclada,* contain 3.2 and 17.3% crude protein, respectively (Penaflorida and Golez, 1996).

Figure 18 indicates that the environmental condition in Patikul, Sulu is conducive to the growth of *H. durvillei* and *U. reticulata* in terms of better proximate composition. For example, in the case of *S. cristaefolium,* Figure 19 indicates that environmental conditions in Burgos, Ilocos Norte can produce a better proximate composition of this species of seaweeds. Furthermore, it has been established that seaweed meal increases fertility and the birthrate of animals and improves the color of salmonids (Chapman and chapman 1980).

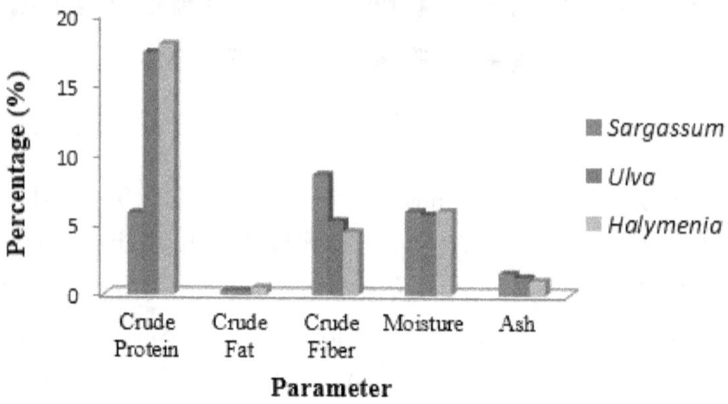

Figure 18. Feed analyses of seaweed species in Patikul, Sulu

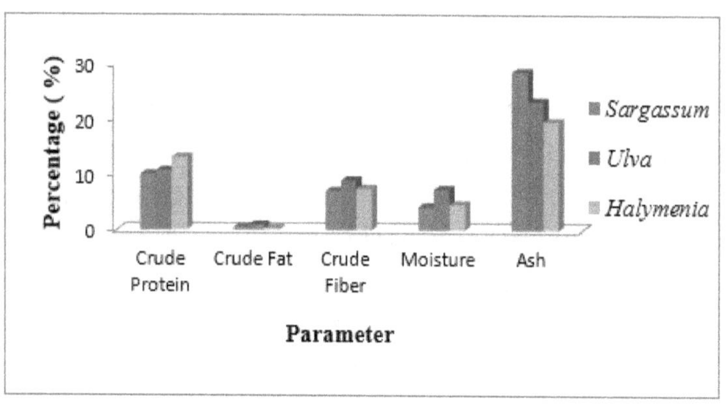

Figure19. Feed analyses of seaweed species in Burgos, Ilocos Norte

There have been several feeding trials conducted to evaluate seaweeds as fish feed. They have been used as fresh and dried meals incorporated as a partial or complete replacement of fishmeal protein in pelleted form. These studies' dietary inclusion levels ranged from 5-100% (FAO,

2009). The addition of *Porphyras pheroplasts* to a semi-purified diet of red seabream (*Marantes corymbosa*) improved their specific growth rate (Kalla et al., 2008). When dried, *Gracilaria busrapastonis* replaced 5 or 10% of the fish protein hydrolysate diet for European seabass (Dicentrarchus labrax) specific growth rate improved (Valente et al. 2006).

Thus, seaweeds evaluated in the present study also have the potential as a source of aquafeed. Feeding the fishes with seaweed helps prevent the occurrence of disease to a certain extent. Seaweeds contain plenty of protein, vitamins, and minerals. Most studies were on tilapia, and submerged macrophytes were fed in fresh form or as dried meals within a pelleted diet.

In addition, 4% of *Sargassum* meal to the feed of shrimp cultures reduced cholesterol contents of their muscle tissue by 29%, quite desirable for shrimp grown for human consumption (Casas- Valdez et al., 2006).

Heavy Metals Uptakes of Seaweed

The aim was to measure concentrations of total copper (Cu), lead (Pb), mercury (Hg), and zinc (Zn) in three edible seaweed species from Burgos, Ilocos Norte, and Patikul, Sulu, Philippines. Copper and Zn are trace minerals essential for life. Copper is a component of many enzymes and proteins required for infant growth, host defense mechanism, bone strength, and more. On the other hand, Zn is involved in cellular metabolism, protein synthesis, a proper sense of taste and smell, and cell division (Sandstead, 2000).

The tolerable upper intake level for Zn is 40 mg but not long-term. Cases of Zn toxicity have been reported. Intakes of 150-450 mg of Zn per day have been associated with chronic effects such as low Cu status, altered iron function, and reduced immune system. Ingesting 4 g of zinc gluconate (570 mg elemental Zn) has caused severe nausea and vomiting (Fosmire, 1990).

Zinc deficiency causes retarded genital, muscular development, and mental lethargy (Maret et al., 2006). Mean dietary Cu intake from adults' food has been estimated in the range of 1.0-2.3 mg/day for males and 0.9-1.8 for females. For omnivore diet, it is 1-1.5 mg/day. Acute Cu toxicity symptoms include salivation, epigastric pain, nausea, vomiting, and diarrhea (Olivares and Uauy, 1998).

Table 5 shows heavy metal concentrations in three seaweeds. Zn registered the highest concentration (328.74 mg/kg) in *U. reticulata* growing in Burgos Ilocos Norte. Other seaweeds registered lower concentrations, and others could be eaten as a source of copper.

Lead and Hg are toxic heavy metals, and they were also detected in seaweeds. *Ulva* seaweeds collected from Burgos, Ilocos Norte, contained a seemingly high level of Pb. However, no maximum contaminant level values specifically set for Pb or Hg in seaweeds were found, so there was no guideline or standard value that could serve as a reference to determine whether the measured concentration is higher or lower than the maximum allowable value.

Table 5. Heavy metal content (mg/kg) of the seaweeds

TAXON	COPPER	LEAD	MERCURY	ZINC	LOCATION
Ulva reticulata	1.75	5.41	0.0199	328.74	Ilocos Norte
	0.0012	0.014	0.0124	0.25	Sulu
Halymenia durvillei	1.70	0.014	0.0106	0.0016	Ilocos Norte
	0.0012	0.035	0.0064	0.53	Sulu
Sargassum cristaefolium	11.45	3.18	0.0414	0.0016	Ilocos Norte
	0.93	0.0 14	0.003	0.0016	Sulu

The concentration of Zn in *U. reticulata* is alarming. Heavy metals of Zn has highest value (328.74 mg/kg) in Burgos, Ilocos Norte compared to Zn (0.25 mg/kg) in Patikul, Sulu. This indicates that these seaweeds are good absorbers of these toxic heavy metals. Considering the results of analyses of these metals elements, U. reticulata from Burgos, Ilocos Norte is not fit for human consumption. Traces of elements like Cu and Zn are beneficial for human intake but should not exceed the tolerable weekly intake suggested by WHO and JECFA (2004). The seaweeds from Patikul, Sulu are safe for human and animal consumptions and cosmetics formulation. Still, in the case of those from Burgos, Ilocos Norte, the concentrations of Zn and Pb in *U. reticulata* and *S. cristaefolium* were to be considered. *Halymenia durvillei* is safe as food, feed, and pharmaceutical preparation.

Seaweeds also have the special property to sequester some pollutants in the water medium, as shown by Troell et al. (1999) when they used seaweed for removing nutrients from intensive mariculture. The present study also shows that seaweeds absorb Pb, Hg, Cu, and Zn, as revealed in the analysis of the three seaweeds species' composition.

Secondary metabolite production of seaweed is a function of life history with different stages having different physiological properties. According to the Journal of Toxicology published in 2009, the single largest source of mercury was ocean fish, but seaweed also contained relatively high levels of these elements. So that seaweed is only typically eaten in a small amount.

In addition, results revealed in this study that heavy metals or trace elements in seaweeds in two different sites (Burgos, Ilocos Norte, and Patikul, Sulu) vary according to the composition of seaweed uptakes of Hg, Pb, Zn, and Cu. This could be due to the effect of location, type of sediments, and salinity of seawater and weather conditions during sampling.

Phytochemical Analyses

Test Tube Method

Phytochemical screening test of three marine algae, namely *Ulva reticulata, Sargassum cristaefolium,* and *Halymenia durvillei* showed the presence of saponins, tannins, and terpenoids in ethanol extract and coumarins, saponins, and terpenoids in methanol extract, as shown in Table 6.

Results of the phytochemical test proved a difference in the capacity of different solvents to extract active compounds. For example, in ethanol extracts, tannins and terpenoids were present in *U. reticulata* (Figures 20 and 21). In Figure 20, reddish-brown precipitated at the interface, showing the presence of terpenoids in ethanol extracts from the three seaweed representatives. In Figure 21, change from brown to brownish green indicates the presence of tannins in *U. reticulata* and Figure 23 shows the presence of saponins in ethanol extract from *H. durvillei*.

Figure 20: The presence of terpenoids in ethanol extract from *Ulva reticulata* *Sargassum cristarfolium* and *Halymenia durvillei*

Both ethanol and methanol seaweed extracts showed that alkaloids, flavonoids, and xanthoprotein were absent in the three seaweed extracts of *U. reticulata, S. cristefolium* and *H. durvillei*. In contrast, the study of Varghese et al. (2010) showed that *U. reticulata* contains alkaloids. The absence of alkoloid in U. reticulata in the present study may be due to the geographical location and seasons of collection. Differences can be attributed to solubility of active component in different solvents (Ekpo and Etim, 2009). Methods of extraction were possible sources of variation for chemical composition and bioactivity of extracts (Felix, 1982). A number of seaweeds from Pakistan coast have been analyzed for their phytochemical investigation (Ahmad et al., 1992). On the other hand, there are very limited information available about phytochemistry of Philippine seaweeds.

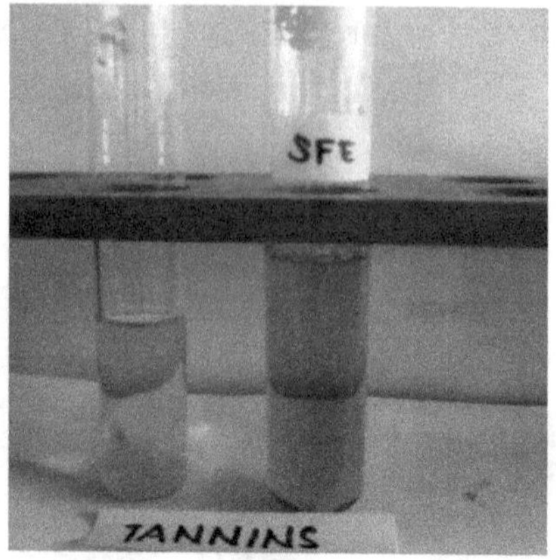

Figure 21. Presence of tannins in ethanol extract from *Ulva reticulata*

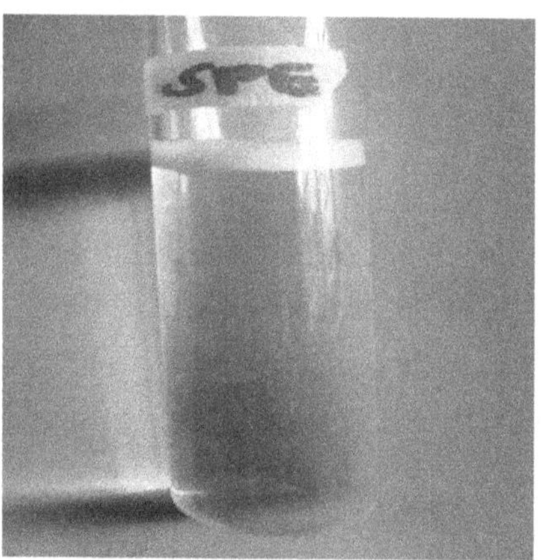

Figure 22. Presence of saponins in ethanol extract from *Halymenia durvillei*

In methanol extracts, *U. reticulata* contains coumarins, saponins and terpenoids. Formation of yellow color indicates presence of coumarins; reddish color denotes

43

presence of terpenoids; and froths formation persisting for ten minutes shows presence of saponins in *U. reticulata* (Figures 23, 24, and 25). Coumarin is used as anticoagulant, drug and in the treatment of lymphoderma.

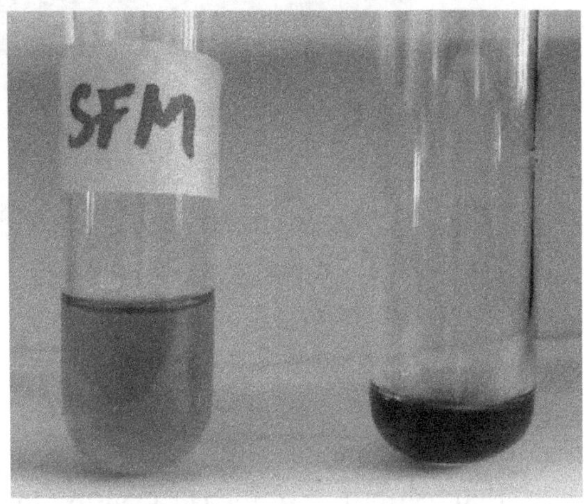

Figure 23. Presence of coumarin in methanol extract of *Ulva reticulata*

Figure 24. Presence of terpenoids in methanol extract of *Ulva reticulata*

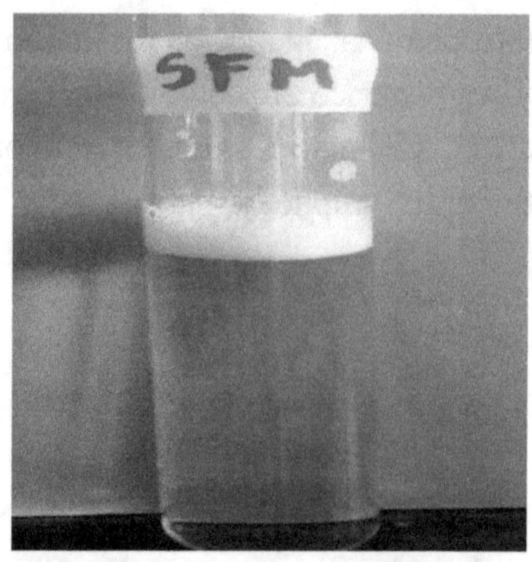

Figure 25. Presence of saponns in methanol extract of *Ulva reticulata*

Also, methanol extracts of *S. cristaefolium* and *H. durvillei* contain terpenoids in average amount, as shown in Figure 26. Reddish-brown at interface shows the presence of terpenoids in two seaweeds species, *S. cristaefolium* and *H. durvillei*

Figure 26. Terpenoids presence in the methanol extract from *Halymenia durvillei* and *Sargassum cristaefolium*

It could be deduced from the screening test that terpenoids are present in the three algal samples, and they may serve as pharmaceutical agents in biomedicine. Terpenoids have found application in the food industry and are antioxidant agents. In addition, antioxidants significantly extend the shelf-life of foods containing oxidizable lipids such as vegetable oils, animal fats, flavorings, spices, nuts, processed meats, and snack products (Finley and Given Jr., 1986).

Terpenoids as pharmaceutical agents act as agents in the formulation of topical drugs. Terpenoids are percutaneous permeation enhancers that increase the solubility of drugs in skin lipid, disruption of lipid/protein organization, and/or extraction of skin micro constituents ' responsible for maintaining barrier status. Hence, terpenoids appear to offer great promise for use in transdermal formulations (Sapra et al., 2008).

There is no doubt that more terpenoids identified from marine algae will become available and will be used in the food industry, and will be utilized as a source of natural antioxidant and clinical drugs that will play significant roles in human disease treatment in the future.

Thin Layer Chromatography

Figures 27 and 28 show the chromatograms of solution A and solution B seaweed extracts. Different extracts of *S. cristaefolium*, *U. reticulata* and *H. durvillei* from Ilocos Norte and Sulu showed steroids, saponins, anthraquinones, phenols, and flavonoids. Chromatogram of solution A extracted with chloroforms: acetic acid and toluene: chloroform developed four spots in each extract. The *S. cristaefolium* from both sites developed four spots each, and the highest spots were recorded in *S. cristaefolium* followed by *U. reticulata* and *H. durvillei* from Sulu. However, only *U. reticulata* had Rf values of 0.41. In contrast, *S. cristaefolium,* and *H. durvillei* from Ilocos Norte had equal Rf values 0.42., while *U. reticulata* from Ilocos Norte had Rf value of 0.41.

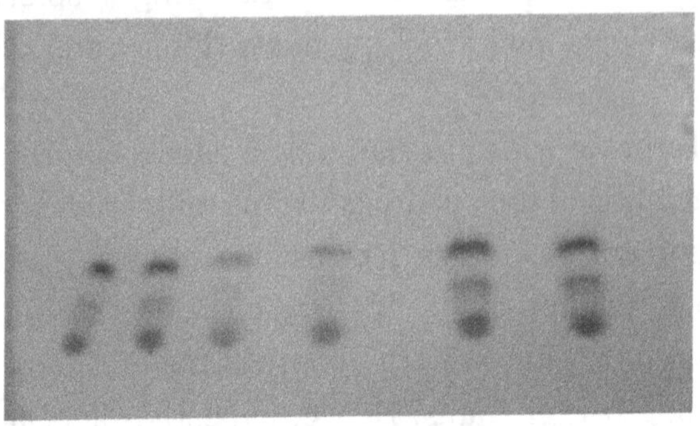

Figure 27. Chromatogram solution A developed using toluene and chloroform

Solution B extracted with methanol: water and developed using chloroform: methanol showed three spots in *S. cristaefolium* with Rf values of 0, 0.18, and 0.20; two spots in *U. reticulata* having Rf values of 0 and 0.11 while three spots in *H. durvillei* with Rf values of 0, 0.22 and 0.24.

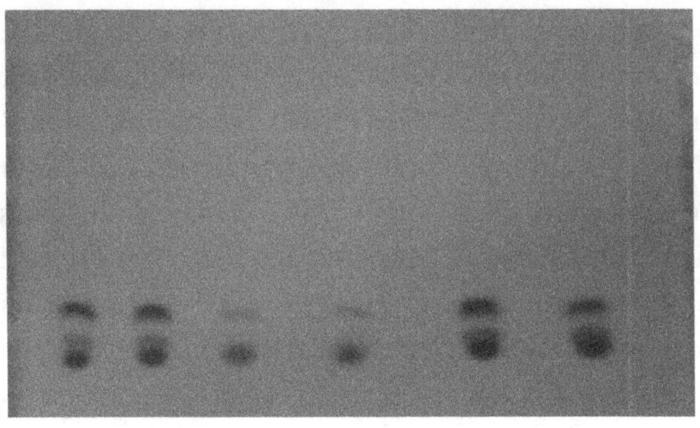

Figure 28. Chromatogram of Solution B developed using chloroform and water

It can be seen that both chromatograms of extracts collected in both sites exhibit similar affinity for both stationary phase and mobile phase, considering similarities in terms of spot location and number of components. This finding indicates that seaweeds have similar phytochemical compositions.

Test for Saponins

Table 6 shows the Rf values of the spots found positive for saponins as these spots formed violet coloration on spraying with vanillin-sulfuric acid (Fig. 29). All the seaweeds collected in both sites contained saponins. Saponin present in *S. cristaefolium* in Sulu has Rf value of 0.20, and that in Ilocos Norte has Rf value of 0.19, which indicates different compositions in terms of structure. For *U. reticulata*, the Rf value of saponin from Sulu is 0.06, while for Ilocos Norte, Rf value of saponin is 0.11, which indicates different compositions in terms of structure. Saponin in Sulu is more polar compared to that of Ilocos Norte. For example, the *H. durvillei* from Sulu has an Rf value of 0.20, while Ilocos Norte has an Rf value of 0.22. This means that saponin from Sulu has a stronger affinity for the stationary phase, which is polar. These secondary metabolites were reported to possess antitumor inflammatory, antiviral, and hemolytic effects and antimutagenic activities and can lower the risk of human cancers by preventing cancer cells from growing. Saponins were present in all seaweeds: *Ulva, Sargassum,* and *Halymenia.* Saponins could be one of the bioactive metabolites responsible for antimicrobial properties.

Table 6. Rf values of saponin in the different seaweeds from two locations

SEAWEED SAMPLE	MEAN RF VALUE	
	Burgos, Ilocos Norte	Patikul, Sulu
Ulva reticulata	0.11	0.06
Halymenia durvillei	0.22	0.20
Sargassum cristaefolium	0.19	0.20

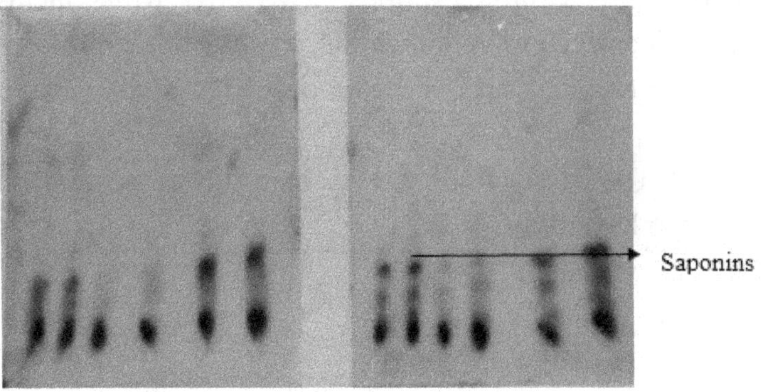

Figure 29. Violet spots indicating presence of saponins

Test for Steroids

Table 7 presents the Rf values of the spots found positive for steroids as these spots formed red-violet coloration upon spraying with vanillin-sulfuric acid (Fig. 30). All seaweeds collected in both sites contained steroids. Steroids present in *S. cristaefolium* in Sulu have Rf of 0.11, and those from Ilocos Norte have similar Rf values of 0.19, which means they could have different steroid compositions. For *U. reticulata*, the Rf value of the steroid from Sulu is 0.11, while for Ilocos Norte, the Rf value of the steroid is 0.12, which indicates different compositions in terms of structure. For example, the steroid in *U. reticulata* growing in Sulu is more polar than that from Ilocos Norte. The steroid structure found in this seaweed has a stronger affinity for the stationary phase, which is polar. The *H. durvillei* of Sulu and Ilocos Norte has the same Rf values of 0.14. Therefore, their steroid structure could be the same.

Steroids, anthraquinones, terpenoids were also noted in the three types of seaweeds studied but could have different structures due to their Rf values. Their metabolites were also reported to exhibit various activities with inclusive applications such as antimicrobial, insecticides, anti-parasitic and cardiotoxic properties (Solomon, et al., 2008).

Table 7. The Rf values of steroids in the different seaweeds from two locations

SEAWEED SAMPLE	MEAN RF VALUE	
	Burgos, Ilocos Norte	Patikul, Sulu
Ulva reticulata	0.12	0.11
Halymenia durvillei	0.14	0.14
Sargassum cristaefolium	0.19	0.11

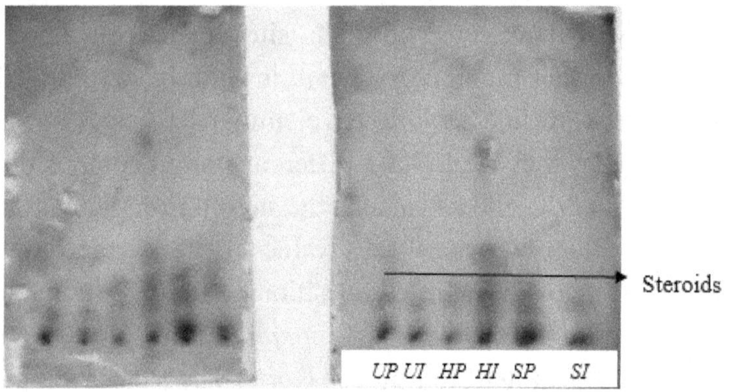

Figure 30. Violet spots showed presence in steroids

Anthraquinones

The Rf values of spots found positive for anthraquinones are shown in Table 8, as these spots formed violet coloration upon spraying with magnesium acetate (Fig. 31). In addition, all the seaweeds collected in both sites contained anthraquinones. Anthraquinones present in *S. cristaefolium* in Sulu has Rf value of 0.41. In contrast, anthraquinones from *S. cristaefolium* from Ilocos Norte has Rf value of 0.42 which means they could have different compositions in terms of structure. For *U. reticulata*, the Rf value of anthraquinones is 0.41 from both sites indicating that they have the same compositions in terms of structure. For example, anthraquinones from *H. durvillei* from Sulu have an Rf value of 0.41, while Ilocos Norte has Rf values of 0.42.

Table 8. The Rf values of anthraquinones from the different seaweed samples

SEAWEED SAMPLE	MEAN RF VALUE	
	Burgos, Ilocos Norte	Patikul, Sulu
Ulva reticulata	0.41	0.41
Halymenia durvillei	0.42	0.41
Sargassum cristaefolium	0.42	0.41

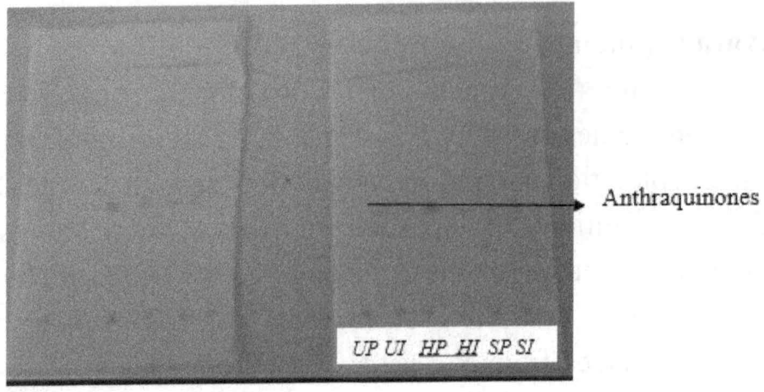

Figure 31. Violet spots indicate presence of the anthraquinones

Flavonoids

The Rf values of flavonoids components of three seaweeds are shown in Table 9. Compounds of the seaweed extracts were detected as yellow spots under UV upon spraying with aluminum chloride (Fig. 32). In addition, all seaweeds collected from Sulu and Ilocos Norte contained flavonoids. Flavonoids present in *S. cristaefolium* in Sulu have Rf value of 0.20 while that from S. *cristaefolium* in Ilocos Norte has Rf value of 0.19 which means they could have different compositions in terms of structure. For *U. reticulata,* the Rf value of flavonoid is 0.19 from both sites, indicating that they have the same composition in terms of structure. Flavonoid from *H. durvillei,* from Sulu, has Rf value of 0.20, while Ilocos Norte has Rf values of 0.22.

Flavonoids have been reported to exert multiple biological properties, including antimicrobial, cytotoxicity, anti-inflammatory, and anti-tumor activities. Still, the best-described property of almost every group of flavonoids is their capacity to act as powerful antioxidants, protecting the human body from free radical and reactive oxygen species (Tapas et al., 2008).

Flavonoids were found present in several species of seaweeds studied. In *U. reticulata,* they are reported to have a wide range of substances that play an important role in protecting biological systems against harmful effects of oxidative processes on macromolecules, such as carbohydrates, proteins, lipids and DNA (Atmani et al., 2009).

Table 9. The Rf values of flavonoids from the different seaweed samples

SEAWEED SAMPLE	MEAN RF VALUE	
	Burgos, Ilocos Norte	Patikul, Sulu
Ulva reticulata	0.19	0.19
Halymenia durvillei	0.22	0.20
Sargassum cristaefolium	0.19	0.20

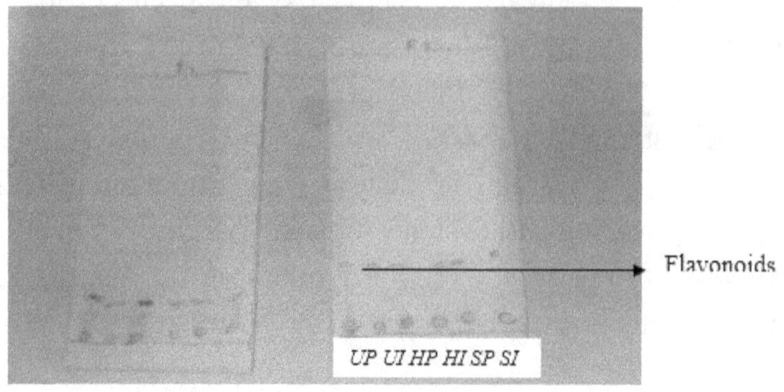

Figure 32. Yellow spots showed presence of flavonoids

Phenols

Eluted compounds of the seaweed extracts were detected as brown spots upon spraying with ferric chloride (Fig. 33). All the seaweeds collected from Sulu and Ilocos Norte were positive for phenols. Phenol present in *S. cristaefolium* in Sulu has Rf value of 0.29 while that from *S. cristaefolium* from Ilocos Norte has Rf value of 0.26 (Table 10) which means that they could have different compositions in terms of structure. For *U. reticulata*, Rf value of the phenol is 0.26 from both sites, indicating that they have the same compositions in terms of structure. Phenols from *H. durvillei* from Sulu have Rf value of 0.27 while those from Ilocos Norte have Rf values of 0.28.

Table 10. Rf values of phenols from the different seaweed samples

SEAWEED SAMPLE	MEAN RF VALUE	
	Burgos, Ilocos Norte	Patikul, Sulu
Ulva reticulata	0.26	0.26
Halymenia durvillei	0.28	0.27
Sargassum cristaefolium	0.26	0.29

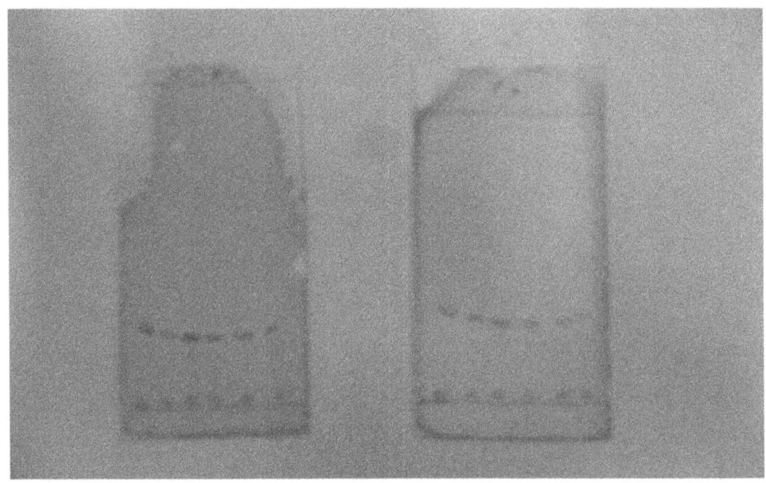

Figure 33. Brown spots showed the presence of phenols

Phenolic acids were reported to increase bile secretion, reduce blood cholesterol and lipid levels, and antimicrobial activity against some strains of bacteria like *Staphylococcus aureus* (Gryglewski, 1987). Tumor cells, including leukaemia cells, typically have higher levels of reactive oxygen species (ROS) than normal cells, so that they are particularly sensitive to oxidative stress (Mandal et al., 2010.)

CHAPTER 6

THE ANTI-BACTERIAL PROPERTY OF SELECTED SEAWEED

Seaweed extract solution showed antibacterial activity against *Staphylococus aureus* and *Escherichia coli*. Zones of inhibition were observed and documented. The *S. cristaefolium* of Ilocos Norte had 5.19±0.52 mm; *S. cristaefolium* of Sulu had 13.35±0.84 mm; and *H. durvillei* of Sulu has 5.98± 1.09 mm zone of inhibition for *E. coli*. *U. reticulata* of Ilocos Norte and Sulu did not affect bacterial pathogens. Maximum activity of *S. cristaefolium* of Ilocos Norte had 4.68± 3.53 mm; and *S. cristaefolium* of Sulu had 15.79 ±0.30 mm inhibition zone against *S. aureus*. *Halymenia durvillei* and *U. reticulata* from both locations have no activity against S. aureus at the time of collection.

Based on the results obtained, *S. cristaefolium* can inhibit E. coli and S. aureus growth. *H. durvillei* can inhibit the growth of *E. coli* but not *S. aureus*. However, *U. reticulata* has no antimicrobial activity against these microbes. In terms of location, extracts from seaweeds in Patikul, Sulu (Figure 34) have stronger antimicrobial activities than extracts from seaweeds in Ilocos Norte (Figure 35) based on their mean zone of inhibition.

The antibacterial activity of seaweeds could be influenced by factors like habitat and season of algal collection. The ability of *S. cristaefolium* to inhibit bacteria is due to secondary metabolites in 10-µl concentration. According to the study of Jena (2008), extract of *Sargassum* sp. was more effective against *B. subtilis* and *E. coli*, showing

18 mm and 16 mm zone of inhibition, respectively, at 400 µg/100 µl concentration in comparison to *S. aureus* showing 10 mm zone of inhibition at the same concentration.

However, in this study, *S. cristaefolium* was more active in *S. aureus* showing 15.79 ±0.84 in 10 µl concentration. Most of the active compounds of marine algae show antibacterial activities. Many metabolites isolated from marine algae have been shown to possess bioactive effects (Vairappan et al., 2011). Therefore, antimicrobial activity depends on algal species and extraction efficiency of their active compounds and location, seasons of the year, and temperature of the water. This difference may be attributed to location or seasonal variations (Perez et al., 1990).

Green seaweed *viz., Caulerpa, Cupvessoides, C. peltata, C. taxifola, Ulva fasciata,* and *U. lactuca* inhibited gram positive S. aureus all the *Vibrio* spp. (Padmakumar, 1998). Moreover, several active principles have been isolated from marine seaweeds, *E. intestinalis* (Ravikumar et al., 2009) and *U. lactuca* (Jothibai et al., 2009). So it is evident that the seaweeds possess antimicrobial compounds.

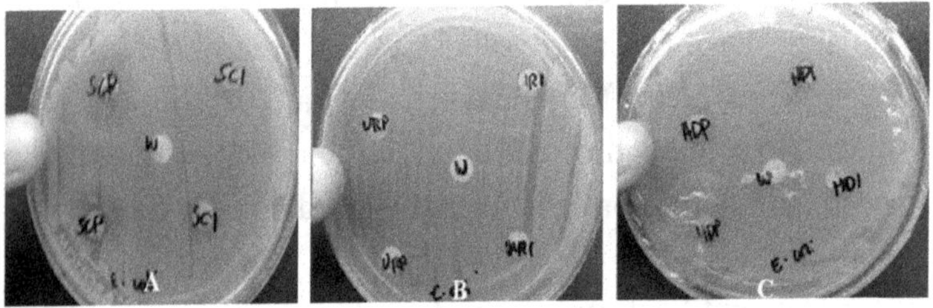

Figure 34. Photograph showing the bacterial plates of *Escherichia coli* inhibited by *Sargassum cristaefolium* Sulu and Ilocos Norte (A) and *Halymenia durvillei* Sulu and Ilocos Norte (C)

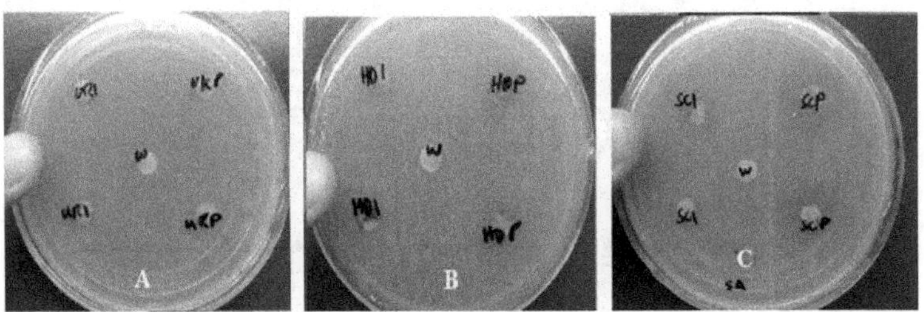

Figure35. Photograph showing the bacterial plates of *Staphylococcus aureus* inhibited by *Sargassum cristaefolium* Sulu and Ilocos Norte (C)

Legend: URP = *Ulva reticulata* Patikul, HDP =*Halymenia durvillei* Patikul

SCP = *Sargassum cristaefolium* Patikul, URI = *Ulva reticulata* Ilocos

HDI = *Halymenia durvillei* Ilocos, SCI = *Sargassum cristaefolium* Ilocos

Standard criteria for the evaluation of seaweed antimicrobial activity are lacking. In some cases, there are significant variations which could be due to the influence of the natural factors of the ecosystem as such: climate, salinity, reproductive state, geographical location, and seasonability in which the seaweed grows. However, other factors like extraction methods and bacteria tested also influence the antimicrobial efficacy of seaweeds. These results of this study show, *S. cristaefolium* of Sulu and Ilocos Norte could inhibit the growth of *E. coli* and *S. aureus*. The *S. aureus,* called golden staph, is a common bacterium that lives on the skin or nose, while *E. coli* attacks the digestive system of the infected individual. The bioactivity of the seaweed extract could be due to the phytochemicals present, which, based on literature, have an antimicrobial effect.

CHAPTER 7

ANTIOXIDANT PROPERTIES OF THE SEAWEEDS

The parameter used to measure free radical scavenging activity of the extract was the percentage radical scavenging activity of diphenyl-p-picrylhydrazyl (% DPPH-RSA). Results of DPPH assay are shown in Table 12. Values were recorded as 119.49 mg/ml, 98.03 mg/ml, 154.73 mg/ml, 225.93 mg/ml, 153.26 mg/ml and 152.91 mg/ml for *S. cristaefolium* of Sulu, *S. cristaefolium* of Ilocos Norte, *U. reticulata* of Sulu, *U. reticulata* of Ilocos Norte, *H. durvillei* of Sulu and *H. durvillei* of Ilocos Norte, respectively. Dried samples of *U. reticulata* of Ilocos Norte showed the best antioxidant activity, while *S. cristaefolium* of Ilocos Norte had the least antioxidant activity.

Figure 36 shows the radical scavenging activity of seaweeds extracts with increasing concentrations. The graph shows an increase in the percentage of free radical scavenging activity of extract with increasing concentration. Among seaweeds, green seaweed, *U. reticulata,* exhibited the highest free radical scavenging or antioxidant activity. Brown seaweed, *S. cristaefolium* had the lowest. Results show that the scavenging activity of seaweeds on the DPPH radical was dependent on concentration.

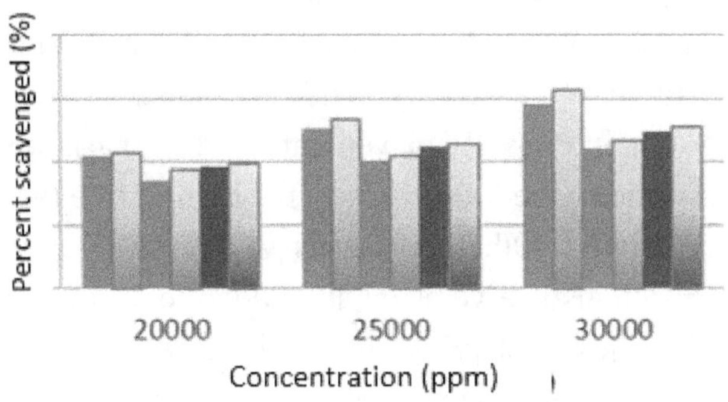

Figure 36. Antioxidant activity (% scavenged) of the seaweeds

The data presented in Table 11 indicates that RSA in extract concentration of 0.50 mg/mL was in the range of 98.03-225.93% for seaweed extracts. Nevertheless, the results of these extracts had higher antioxidant activity than the commercial antioxidant ascorbic acid (100 ppm, 89.83%).

Table 11. Radical Scavenging Activity in the extract concentration

SEAWEED SAMPLE	MEAN RF VALUE DPPH ASSAY IC50 (mg/ml)	
	Burgos, Ilocos Norte	Patikul, Sulu
Ulva reticulata	225.93	154.73
Halymenia durvillei	152.91	153.26
Sargassum cristaefolium	98.03	119.49

Among species, *U. reticulata* is a strong potential source of antioxidants. This seaweed has the potential of scavenging radicals caused by oxidative stress. Different results were obtained by Raghavendran et al. (2004), where *Sargassum* species were found to have the highest free radical scavenging activity. Other seaweeds *H. durvillei* and *S. cristaefolium* can also serve as an antioxidant only to a little degree. With antioxidant properties of these seaweeds, they can be applied in pharmaceuticals to prevent free-radical induced diseases like cancer. The antioxidant property of the seaweeds can be attributed to phenolic metabolites such as flavonoids, saponins, terpenoids, and tannins. Seaweed is one of the potential objects for extracting anti-inflammatory agents (Kijjoa, 2004). Like *Ulva lactuca* the green alga available in Tuticorin coast showed an anti-inflammatory effect (Jothibai et al., 2009).

LITERATURE CITED

AGNGARAYNGAY, Z.M. 1983. Species of edible seaweeds of Ilocos Norte. Ilocos Fishery Journal 1(2): 118-133.

AGNGARAYNGAY, Z.M. 2000. Edible Seaweeds of Ilocos Norte: Food Preparations other Local Uses and Market Potentials.

AHMAD, H., M. SURIF, W.M.W. OMAR, M.N. BIN ROSLI, and M.D. NOR AR. 2011. Nutrient uptake, growth and chlorophyll content of green seaweed, *Ulva reticulata*: Response to different source of inorganic nutrients. In: Universiti Malaysia Terengganu 10th International Annual Symposium, Kuala Terengganu, Malaysia, 11-13 July 2011. Kuala Terengganu, Malaysia: Universiti Malaysia Terengganu.

ALVEAL, K. 1998. The seaweed resources of Chile. In: A.T. Crithley and M. Ohno (eds). Seaweed Resources of the World. Japan International Cooperation Agency, Yokosuka: 347-365

ANANTHARAMAN P., K. MANIVANNAN, G. KARTHIKAIDEVI and G. THIRUMARAN. 2009. Element composition of certain seaweeds from Gulf of Mannar Marine Biosphere Reserve, Southeast Coast of India. World Journal of Dairy and Food Sciences 4 (1):46-55.

AOAC. 1990. Official Methods of Analysis, 15th edition. Association of Official Analytical Chemists Washington D.C., USA.

APOSTOLIDIS, E., and C.M. LEE. 2010. In-vitro potential of *Ascophyllum nodosum* phenolic antioxidant-mediated alpha-glucosidase and alpha-amylase inhibition. Journal Food Science 75(3): 97-102.

ATEWEBERHAN, M., R. PRUD'HOMME, and W.F. VAN. 2005. A taxonomic survey of seaweeds from Eritrea. Blumea 50: 65-111.

ATHUKORALA, Y., K.N. KIM, and Y.J. JEON. 2006. Anti-proliferative and antioxidant properties of an enzymatic hydrolysate from brown alga, *Ecklonia cava*. Food Chemistry Toxicology 44: 1065-1074.

ATMANI, D., C. NASSIMA, A. DINA, B. MERIEM, D. NADJET, and B. HANIA. 2009. Flavonoids in human health: from structure to biological activity. Current Nutrition and Food Science 5: 225-237.

BENTLEY, K.W. 1957. The alkaloids: Inter-science: New York, NY, USA. p. 1.

BHAT, S.V., B.A. NAGASAMPAGI, and M. SIVAKUMAR. 2005. Chemistry of Natural Products. Narosa Publishing House, India.

BLIGH, E.G., and W.J. DYER. 1959. A rapid method of total lipid extraction and purification. Canadian Journal of Biochemistry and Physiology 37:913–917.

BRAVO, L. 1998. Polyphenols: Chemistry, dietary sources, metabolism, and nutritional significance. Nutritional Revised 56: 317-333.

BRIONES, A.V., W.O. AMBAL, R.R. ESTRELLA, R. PANGILINAN, C.J. DE VERA, and R.L. PACIS. 2004. Tensile and tear strength of carrageenan film from Philippine *Eucheuma* species. Marine Biotechnology 6:148– 151.

BRITISH NUTRITION FOUNDATION. 2007. Nutrient requirement and recommendation. Available from http://www.britishnutrition.org. Accessed November 2007.

CASAS-VALDEZ, M., H. HERNANDEZ-CONTRERAS, A. MARIN-ALVAREZ, R. AGUILA-RAMIREZ, C.J. HERNANDEZ-GUERRERO, L. SANCHEZ- RODRIGUEZ, and S. CARILLO-DOMINGUEZ. 2006. The seaweed *Sargassum* (Sargassaceae) as tropical alternative for goats' feeding. Revista de Biologia Tropical 54(1): 83-92.

CAVALIER-SMITH, T. 2007. Evolution and Relationships of Algae: Major Branches of the Tree of Life. pp 21-55. In: J. Brodie and J. Lewis (eds). Unravelling the Algae: the Past, Present and Future of Algal Systematics, CRC Press, Boca Raton, London and New York.

CHANDA S., R. DAVE, M. KANERIA, and K. NAGANI. 2010. Seaweeds: a novel, untapped source of drugs from sea to combat infectious diseases. In: A. Mendez-Vilas (ed). Current Research, Technology and Education Topics in Applied Microbial Biotechnology. pp. 473-480.

CHAPMAN, V. J. AND CHAPMAN, D. J. 1980. Seaweed and Their Uses. Chapman & Hall, London.

CHENNUBHOTLA, V.S.K, and N. KALIAPERUMAL. 1998. Seaweeds and their importance. Proc. of the Workshop National Aquaculture Week. Published by the Aquaculture Foundation of India, Chennai. pp. 54 – 57.

CHOPIN, T. 2007. Closing remarks of the new president of the International Seaweed Association. In: The 19th International Seaweed Symposium, March 26-31, 2007 Kobe, Japan. pp 1-2.

COPPEJANS, E., F. LELIAERT, O. DARGENT, R. GUNASEKARA, and O.D.E. CLERCK. 2009. Sri Lankan Seaweeds. Methodologies and Field Guide to the Dominant Species. In: S. Yves, D.V. Spiegel and J. Degreef (eds). ABC Taxa. A Series of Manuals Dedicated to Capacity Building in Taxonomy and Collection Management. Belgian Development Cooperation.

COSMETIC INGREDIENTS DICTIONARY (2002-2011) Algae http://www.cosmeticscop.com/cosmeticingredients-dictionary/A.aspx cited16 Dec.2011.

COX, S., N. GHANNAM, and S. GUPTA. 2010. An assessment of the antioxidant and antimicrobial activity of six species of edible Irish seaweeds. International Food Research Journal 17: 205-220.

CRESPY, V. and G. WILLIAMSON. 2004. A review of the health effects of green tea catechins: in vivo animal models. Journal of Nutrition 134: 3431S-3440S.

CUSHNIE, T., B. CUSHNIE, and A. LAMB. 2014. Alkaloids: an overview of their antibacterial, antibiotic-enhancing and anti-virulence activities. International Journal Antimicrobial Agents 44(5): 377–386. Retrieved on September 2014 from doi:10.1016/j.ijantimicag.2014.06.001. PMID25130096.

DAGOON, J. and P.D. SANGATANAN. 2005. Physical Education, Health, and Music I. Philippines: Rex Book Store, Inc.

DARCY-VRILLÓN, B. 1993. Nutritional aspects of the developing use of marine macroalgae for the human food industry. International Journal on Food Science and Nutrition 44: 23-35.

DAVID, M.J. and R.G. PAREKH, 1975. Protein content of green seaweeds from the Sourashtra coast. Salt Research Industry 11(2): 41-44.

DAWES, C J. 1998. Marine Botany. John Wiley and Sons, Inc., New York, P.480

DEMBITSKY, V.M., H. REZANKOVA, T. REZANKA, and L.O. HANUS. 2003. Variability of the fatty acids of the marine green algae belonging to the genus *Codium*. Biochemistry System Ecology 31:1125–1145.

DHARGALKAR, V.K. and D. KAVLEKAR. 2004. Seaweeds - a Field Manual. (1st Edn). National Institute of Oceanography, Goa. pp 1-2.

DONE, T.J. 1992. Phase shifts in coral reef communities and their ecological Significance. Hydrobiologia 247:121-132.

EISLER R. 2000. Handbook of Chemical Risk Assessment-Health Hazards to Humans, Plants and Animals-Metals: Florida, USA, Lewis Publishers, Boca Raton.

EKPO M.A. and P.C. ETIM. 2009. Antimicrobial activity of ethanolic and aqueous extracts of *Sida acuta* on microorganisms from skin infections. Journal of Medicinal Plants Research 3: 621- 624.

ETCHEVERRY, D.H. 1960. Algas marinas de las Oceanas chilenas (Juan Fernandez, San Felix. San Ambrosio y Pascua). Revista de Biologia Marina 10: 83-138.

FAO 2009. The state of world fisheries and aquaculture, 2008. FAO Fisheries and 531 Aquaculture Department. Food and Agriculture Organization of the United Nations, 532 Rome.

FARINOLA, N. and N. PILLER. 2005. Pharmacogenomics: its role in re-establishing coumarin as treatment for lymphedema. Lymphatic Research and Biology 3 (2): 81–86. Retrieved on September 2014 from doi:10.1089/lrb.2005.3.81. PMID16000056.

FIGUEIRA, M.M., B. VOLESKY, V.S.T. CIMINELLI, and F.A. RODDICK. 2000. Biosorption of metals in brown seaweed biomass. Water Resources 34:196–204

FELIX M.T. 1982. Medicinal Microbiology. Churchill Livingstone: London, U.K. 848 pp.

FINLEY, J.W. and P. GIVEN, JR. 1986. Technological necessity of antioxidants in the food industry. Food and Chemical Toxicology 2(10-11): 997-1255.

FLEURENCE, J. 1999. Seaweed proteins: biochemical, nutritional aspects and potential uses. Trends Food Science Technology 10:25–38.

FOOD SAFETY AUTHORITY OF IRELAND. 1999. Recommended dietary allowances for Ireland Dublin.

FOSMIRE, G.J. 1990. Zinc Toxicity. American Journal Clinical Nutrition 51(2): 225- 227

FOUREST, E. and B. VOLESKY. 1997. Alginate properties and heavy metal biosorption by marine algae. Applied Biochemistry and Biotechnology 67:33–44.

GEORGE, F., H. ZOHAR, P. HARINDER, and B. KLAUS. 2002. The biological action of saponins in animal systems: a review. Britannica Journal on Nutrition 88(6): 587-605.

GEORGE, T.W., E. PATERSON, S. WAROONPHAN, M.H. GORDON and J.A. LOVEGROVE. 2012. Effects of chronic consumption of fruit and vegetable puree-based drinks on vasodilation, plasma oxidative stability and antioxidant status. Journal on Human Nutrition and Dietetics 25(5): 477-487.

GRAHAM, L.E. and L.W. WILCOX. 2000. Algae. New Jersey, USA. Prentice-Hall, Upper Saddle River, N.J. 640 pp.

GRYGLEWSKI, R.J., R. KORBUT, and J. ROBAK. 1987. On the mechanism of antithrombotic action of flavonoids. Biochemical Pharmacology 36: 317-321.

GUEVARRA, B. 2005. A Guidebook to Phytochemical Screening: Phytochemical and Biological. Revised edition. Manila: UST Publishing House. p. 156

GUILLERMO, D.P. and L.J. MCCOOK. 2008. Environmental status: macroalgae (Seaweeds).

GUIRY, M.D. and C.C. HESSION. 1998. The seaweed resources of Ireland. In: A.T. Critchley and M. Ohno (eds). Seaweed Resources of the World Japan International Cooperation Agency, Yokosuka. pp. 210-216.

GUIRY M.D., J. MORRISSEY, and S. KRAAN. 2001. A Guide to Commercially Important Seaweeds on the Irish Coast Dublin. Irish Fisheries Board. pp. 5–40.

GUISSEPPE, R. and T.M. BARATTA. 2000. Antioxidant activity of selected essential oil components in two lipid model systems. African Journal Biotechnology 69(2): 167-174.

HAHN, D. E. 2000. Applied thin-layer chromatography. WILEY-VCH.

HANAA, H., E. ABD, L. BAKY, K. FAROUK, E.L. BAZ, S. GAMAL and E.L. BAROTY, 2008. Evaluation of marine alga *Ulva lactuca* as a source of natural preservative ingredient, Journal of Agriculture and Environmental Science 3: 434-444.

HANSON, J.R. 2003. Natural products the secondary metabolites. The Royal Society of Chemistry 1-27. Cambridge, UK.

HARBORNE, J.B. 1998. Phytochemical Methods: a Guide to Modern Techniques of Plant Analysis. Chapman and Hall, London, Great Britain.

HARBORNE, J.B. and C.A. WILLIAMS. 2000. Advances in flavonoid research since 1992. Phytochemistry 55: 481-504.

HARDER, T. and P.Y. QIAN. 2000. Waterborne compounds from the green seaweed *Ulva reticulata* as inhibitive cues for larval attachment and metamorphosis in the polychaete *Hydroides elegans*. Biofouling 16:205-214.

HARDER, T., S. DOBRETSOV, and P.Y. QIAN. 2004. Waterborne polar macromolecules act as algal antifoulants in the seaweed *Ulva reticulata*. Marine Ecology Program Service 27: 133-141.

HERBRETEAU, F., L.J.M. COIFFARD, A. DERRIEN, and Y. DE ROECK-HOLTZHAUER. 1997. The fatty acid composition of five species of macroalgae. Botanica Marina 40, 25–27.

HOUGHTON, P.J., A.Y. MENSAH, N. IESSA, and L.Y. HONG. 2003. Terpenoids in Buddleja: relevance to chemosystematics, chemical ecology and biological activity. Phytochemistry 64: 385-393.

JOTHIBAI, R., S. MARGRET, S. KUMARESAN and S. RAVIKUMAR, 2009. A preliminary study on the anti-inflammatory activity of methanol extract of *Ulva lactuca* on rat. Journal of Environmental Biology 30(5): 899–902.

Joint FAO/WHO Expert Committee on Food Additives (JECFA) 63rd Meeting. 2004

KADAM, S.U. and P. PRABHASANKAR 2010. Marine foods as functional ingredients in bakery and pasta products. Food Research International 43: 1975-1980.

KARTHIKAIDEVI, G., K. MANIVANNAN, G. THIRUMARAN, P. ANANTHARAMAN and T. BALASUBARAMANIAN. 2009. Antibacterial properties of selected green seaweeds from Vedalai Coastal Waters: Gulf of Mannar Marine Biosphere Reserve. Global Journal of Pharmacology 3(2): 107-112.

KALLA, A., T. YOSHIMATSU, T. ARAKI, D. ZHANG, T. YAMAMOTO, and S. SAKAMOTO. 2008. Use of *Porphyra* spheroplasts as feed additive for red sea bream. Fisheries Science 74: 104-108.

KAWAGUCHI, S., S. SHIMADA, S. ABE, and R. TERADA. 2005. Morphological and molecular phylogenetic studies of a red alga, *Halymenia durvillei,* (Halymeniaceae, Halymeniales) from Indo-Pacific. Coastal Marine Science 30(1): 201-208.

KHAIRY, H.M. and S.M. EL-SHAFAY. 2013. Seasonal variations in the biochemical composition of some common seaweed species from the Coast of Abu Qir Bay, Alexandria, Egypt. Oceanologia 55: 435-452.

KIJJOA, A. and P. SAWANGWONG. 2004. Drugs and cosmetics from the sea. Marine Drugs 2: 73-82.

KITTAKOOP, P., C. MAHIDOL, and S. RUCHIRAWAT. 2014. Alkaloids as important scaffolds in therapeutic drugs for the treatments of cancer, tuberculosis, and smoking cessation. Current Top Medical Chemistry 14 (2): 239.

KOLANJINATHAN, K. and D. STELLA. 2011. Comparitive studies on antimicrobial activity of *Ulva reticulata* and *Ulva lactuca* against human pathogens. International Journal of Pharmaceutical and Biological Archives 2(6):1738-1744.

KULLING, S.E. and H.M. RAWEL. 2008. Chokeberry (*Aronia melanocarpa*) – A review on the characteristic components and potential health effects. Planta Medica 74: 1625-1634.

LAPASIN, R. and S. PRICI. 1995. Rheology of Industrial Polysacharides – Theory and Application, London, Blackie Academic and Professional.

LEDNICER, D. 2011. Steroid Chemistry at a Glance. Hoboken: Wiley.

LEWMANOMONT, K. 1998. The Seaweed Resources of Thailand. In: A.T.Critchley and M. Ohno (eds). Seaweed Resources of the World. Yokosuba, Japan: JICA, pp. 70-78.

LI, H., Z. WANG and Y. LIU. 2003. Review in the studies on tannins activity of cancer prevention and anti-cancer. Zhong Yau Cai 26: 444-448.

LIU, H. 2011. Extraction and Isolation of Compounds from Herbal Medicines. In: J. Willow and H. Liu (eds.) Traditional Herbal Medicine Research Methods. John Wiley and Sons, Inc.

MANIVANNAN K., G. THIRUMARAN, G. KARTHIKAI DEVI, A. HEMALATHA, and P. ANANTHARAMAN. 2008.

Biochemical composition of seaweeds from Mandapam coastal regions along southeast coast of India. American - Eurasian Journal of Botany 1(2): 32-37

MARINHO-SORIANO E., P.C. FONSECA. M.A.A. CARNEIRO, W.S.C. MOREIRA WSC. 2006. Seasonal variation in the chemical composition of two tropical seaweeds. Bio-resources Technology 97:2402-2406.

MCCANCER, E.M. and H.B. WIDDOWSON. 1993. McCance and Widdowson's Composition of Foods. 6th ed. Cambridge. Royal Society of Chemistry.

MCCOOK, L.J. 1999. Macroalgae, nutrients and phase shifts on coral reefs. Scientific issues and management consequences for the Great Barrier Reef. Coral Reefs 18: 357-367

MICHAEL, T.M., M.M. JOHN, and P. JACK. 2005. Brock microbiology of microorganisms. 11th Edition, New Jersey.

MICHAEL, C.H., 2008. Western Poison-Oak, *Toxicodendron diversilobum*. 1st Edition Global Twitcher. Nicklas Stromberg. p. 49.

MOHANTY, D., S.P. ADHIKARY, and G.N. CHATTOPADHYAY. 2013. Seaweed liquid fertilizer (SLF) and its role in agriculture productivity. The Ecoscan Special issue 3: 147-155.

MONTANO, M.N., M.C. RODRIQUEZA, and R.L. BALITAAN. 2006. Ethnobotany of *Sargassum* spp. in the Philippines. Coastal Marine Science Journal 30(1): 222-225.

MORGAN, H., B.H.L. CHUA, E.O. FULLER, and D. SIEHL. 1980. Regulation of protein synthesis and degradation during in *vitro* cardiac work. American Journal of Physiology 238: E431-E442.

MSUYA, F.E. and A. NEORI. 2002. *Ulva reticulata* and *Gracilaria crassa*: macroalgae that can biofilter effluent from tidal fishponds in Tanzania. Western Indian Ocean Journal of Marine Science 1: 117-216.

MUNIER, M., J. DUMAY, M. MORANÇAIS, P. JAOUEN, and J. FLEURENCE. 2013. Variation in the biochemical composition of the edible seaweed *Grateloupia turuturu* Yamada harvested from two sampling sites on the Brittany Coast (France): the influence of storage method on the extraction of the seaweed pigment R-phycoerythrin. Journal of Chemistry, Article ID: 568548, 8 pp. http://dx.doi.org/10.1155/2013/568548

NANG, H.Q. and N.H. DINH. 1998. The Seaweed Resources of Vietnam. In: A.T. Critchley and M. Ohno (eds), Seaweed Resources of the World. Yokosuba, Japan: JICA, pp. 62-69.

NISIZAWA, K., H. NODA, R. KIKUCHI and T. WATAMABA. 1987. The main seaweeds food in Japan. Hydrobiologia 151/152: 5–29.

NORTON, T.A., M. MELKONIAN and R.A. ANDERSON 1996. Algal biodiversity. Phycologia 35: 308-326.

OHNO, M. (ed). 1998. Seaweed Resources of the World. Kanegawa International Fisheries Training Centre, JICA, Yokosuka, Japan, pp. 47-61.

ONOFREJOVA, L., J.V. VASICKOVA, B. KLEJDUS, P. STRATIL, L. MISURCOVA, S. KRACMAR, J. KOPECKY, and J. VACEK. 2010. Bioactive phenols in algae: the application of pressurized-liquid and solid-phase extraction techniques. Journal of Pharmaceutical and Biomedical Analysis 51: 464-470.

PADMA KUMAR, K. 1988. Bioactive substances from marine algae and mangroves. Ph. D. thesis. Annamalai University, Tamil Nadu, India.

PÁDUA, M., P.S.G. FONTOURA, and A.L. MATHIAS. 2004. Chemical composition of *Ulvaria oxysperma* (Kützing) Bliding, *Ulva lactuca* (Linnaeus) and *Ulva fasciata* (Delile). Brazilian Archives of Biology and Technology 47: 49-55

PAGDILAO, C.R. 2002. Seaweed projects funded by PCAMRD. In: A.Q.Hurtado, N.G. Guanson, Jr., T.R. de Castro- Mallare and M.R.J. Luhan (eds.) Proceedings of the National Seaweed Planning Workshop held on August 2-3, 2001, SEAFDEC Aquaculture Department.

PARTHIBAN C., K. PARAMESWARI, C. SARANYA, A. HEMALATHA, and P. ANANTHARAMAN. 2012. Production of sodium alginate from selected seaweeds and their physiochemical and biochemical properties. Asian Pacific Journal of Tropical Biomedicine 2: 1-4.

PHANG, S.M. 1998. The Seaweed Resources of Malaysia. In: A.T. Critchley and M. Ohno (eds.). Seaweed Resources of the World. Yokosuka, Japan: Japan International Cooperation Agency. pp. 79-91.

PEARSON, N.J., C.F. CULLEN, N.S. DZHINDZHEV and H. OHKURA. 2005. A pre-anaphase role for a Cks/Suc1 in acentrosomal spindle formation of *Drosophila* female meiosis. EMBO Reports 6(11): 1058-63.

PEARSON, D. 1999. Pearson's Composition and Analysis of Foods. University of Reading.

PENAFLORIDA, V.D. and N.V. GOLEZ. 1996. Use of seaweed meals from *Kappaphycus alvarezii* and *Gracilaria heteroclada* as binders in diets for juvenile shrimp *Penaeus monodon*. Aquaculture 143(3-4): 393-401.

PEREZ, R.M., J.G. AVILA, and G. PEREZ. 1990. Antimicrobial activity of some American algae. Journal of Ethno- pharmacology 29: 111-118

PHILLIPS, G.O. and P.A. WILLIAMS (eds). 2000. Handbook of Hydrocolloids, CRC Press, ISBN-978 1 84569 414 2, Boca Raton, Florida.

PISE N. M. and A. B. SABALE. 2010. Biochemical Composition of Seaweeds along Central West Coast of India. Pharmacognosy Journal 2(7): 148-150.

RADWAN, A. and A. SALAMA. 2006. Market basket survey for some heavy metals in Egyptian fruits and vegetables. Food and Chemical Toxicology 44: 1273-1278.

RAGHAVENDRAN, H.R.B., A. SATHIVEL, and T. DEVAKI. 2004. Hepatoprotective nature of seaweed alcoholic xtract on acetaminophen induced hepatic oxidant stress. Journal of Health, 50(1): 42-46

RANDHIR, R., Y.T. LIN, and K. SHETTY. 2004. Phenolics, their antioxidant and antimicrobial activity in dark germinated fenugreek sprouts in response to peptide and phytochemical elicitors. Asia Pacific Journal on Clinical Nutrition 13: 295-307.

RAO, D.V. and B. RAO. 2004. Drugs from marine algae - current status, Proceeding Symposium Seaweed 2004. Seaweed Research and Utilization Association and Central Marine Fisheries Research Institute, Cochin India. p. 54.

RATANA-ARPORN P. and A. CHIRAPART. 2006. Nutritional evaluation of tropical green seaweeds *Caulerpa lentillifera* and *Ulva reticulata*, Kasetsart Journal on Natural Science 40: 75–83.

RAYMOND S.J., S. JAHK, and J.M. PITCHFORD. 2010. The Essence of Analgesia and Analgesics. Cambridge University Press. pp. 82–90. ISBN 1139491989.

RAVIKUMAR, S., G. RAMANATHAN, M. SUBHAKARAN, and S. JACOB INBANESON. 2009. Antimicrobial compounds from marine halophytes for silkworm disease treatment. International Journal of Medicine and Medical Sciences 1(5):184-190.

RHEN, T, and J.A. CIDLOWSKI. 2005. Anti-inflammatory action of glucocorticoids new mechanisms for old drugs. English Journal on Medicine 353(16): 117-123. PMID 16236742.

REETA, J. 1993. Seasonal variation in biochemical constituents of *Sargassum wighti* (Grevillie) with reference to yield in alginic acid content. Seaweed Research and
Utilisation 16 (1-2): 13-16.

ROMERO, J.B. 2002. Seaweed farming in the Sulu archipelago. pp. 15-21. In: A.Q. Hurtado, N.G. Guanzon, Jr., T.R. de Castro-Mallare, and M.R.J. Luhan (eds) Proceedings of the National Seaweed Planning Workshop held on August 2-3, 2001, SEAFDEC Aquaculture Department, Tigbauan, Iloilo.

RUSSO, P., A. FRUSTACI, A. DEL BUFALO, M. FINI, and A. CESARIO. 2013. Multitarget drugs of plants origin acting on Alzheimer's disease. Current Medical Chemistry 20 (13): 1686-93. PMID 23410167.

SANCHEZ-MACHADO, D.I., J. LOPEZ-HERNANDEZ, P. PASEIRO-LOSADA, and J. LOPEZ-CERVANTES. 2004. An HPLC method for the quantification of sterols in edible seaweeds. Biomedical Chromatography 18: 183-190.

SANCHEZ-RODRIGUEZ, I., M.A. HUERTA-DIAZ., E. CHOUMILINE., O. HOLGUIN-QUINONES and J.A. ZERTUCHE- GONZALEZ. 2001. Elemental concentrations in different species of seaweeds from Loreto Bay, Baja California Sur, Mexico: implications for the geochemical control of metals in algal tissue. Environmental Pollution 114: 145–160.

SANDSTEAD, H.H. and L.M. CLEAVY. 2000. History of nutrition symposium: Trace element nutrition in human health. Nutrion 130: 483S-484S.

SAPRA, B., S. JAIN, and A.K. TIWARY. 2008. Percutaneous permeation enhancement by terpenes: Mechanistic View. The AAPS Journal 10(1): 120-132.

SARKER, S.D., Z. LATIF and A.I. GRAY. 2006. Natural Products Isolation, 2nd Ed, Humana Press, New Jersey, United States of America.

SARKER, S.D. and L. NAHAR. 2005. Hyphenated techniques. In: S.D. Sasker, Z. Latif, and A.T. Gray (Eds). Natural Products Isolation, 2nd Ed. Totowa, Humana Press.

SCHAFFELKE, B, J. MELLORS, and N.C. DUKE. 2005. Water quality in the Great Barrier Reef region: responses of mangrove, seagrass and macroalgal communities. Marine Pollution Bulletin 51: 279-296.

SCHALBERT, A. 1991. Antimicrobial properties of tannins. Phytochemistry 30(12): 3875-3883.

SCHIEWER S. and B. VOLESKY. 1999. Biosorption by marine algae. In: Remediation, J. J. Valdes, A. Kluwer, and C. Dordrecht (eds). The Netherlands.

SCHRIPSEMA, J., A. RAMOS-VALDIVIA, and R. VERPOORTE. 1999. Robustaquinones, novel anthraquinones from elicited *Cinchona robusta* suspension culture. Phytochemistry 51: 55-60.

SEBASTIANI, L., F. SCEBBA F. and T.R. TONGE. 2004. Heavy metal accumulation and growth responses in poplar clones eridano (*Populus deltoids* × *Maximo wiczii)* and I-214 *(Populus* × *Euramericana)* exposed to industrial waste. Environmental Exploratory Botany 52: 79.

SHANMUGAM, A.N.D. and C. PALPANDI. 2008. Biochemical composition and fatty acid profile of the green algae *Ulva reticulata*. Asian Journal of Biochemistry 3: 26-31.

SHIMADA, K., K. FUJIKAWA, K. YAHARA, and T. NAKAMURA. 1992. Antioxidative properties of xanthone on the autooxidation of soybean in cylcodextrin emulsion. Journal of Agricultural and Food Chemistry 40: 945-948.

SHIN, H.C., H.J. HWANG, K.J. KANG and B.H. LEE. 2006. An antioxidative and anti-inflammatory agent for potential treatment of osteoarthritis from Ecklonia cava. Archives of Pharmaceutical Research 29(2): 165-171.

SMIT, A.J. 2004. Medicinal and pharmaceutical uses of seaweed natural products. A review. Journal Applied Phycology 16: 245-262.

SOLOMON, R.D.J. and V.S. SANTHI. 2008. Purification of bioactive natural product against human microbial pathogens from marine seaweed *Dictyota acutiloba*, Journal of Microbiological Biotechnology 24: 1747- 1752.

TAPAS, A.R., D.M. SAKARKAR, and R.B. KAKDE. 2008. Flavonoids as nutraceuticals: a review. Tropical Journal of Pharmaceutical Research 7: 1089-1099.

TOBIN, M.L., S. MARSHAM and G.W. SCOTT. 2007. Comparison of nutritive chemistry of a range of temperate seaweeds. Food Chemistry 100: 1331–1336.

TREASE, G.E. and W.C. EVANS. 2002. Pharmacognosy. 15th Ed. London: Saunders Publishers; pp. 42–44, 221–229, 246–249, 304–306, 331–332, 391–393.

TROELL, M., P. RÖNNBÄCK, C. HALLING, N. KAUTSKY, and A. BUSCHMANN. 1999. Ecological engineering in aquaculture: use of seaweeds for removing nutrients from intensive mariculture. Journal of Applied Phycology 11: 89–97

TRONO, G.C. JR. 2001. Seaweeds: FAO Species Identification Guide for Fishery Purposes. FAO, Rome. 686 pp.

TRONO, G.C. JR. 1993. Environmental Effects of Seaweed Farming. SICEN Newsletter 4(1): 1-7.

TRONO, G.C. JR., 1992. Preliminary report on the site evaluation/ assessment and collection trip for seaweed farming at Sorsogon, the project area. Seaweed Production Development Project. PHI/89/004 BFAR/UNDP/FAO Philippines. Field Document 06.

TRONO, G.C. JR. and E.T. GANZON-FORTES. 1988. Philippines Seaweeds. Technology and Livelihood Resource Center. National Book Store, Inc. Metro Manila, Philippines. 30 pp.

UAUY, R., M. OLIVARES, and M. GONZALEZ. 1998. Essentiality of copper in humans. American Journal on Clinical Nutrition 67(suppl): 952S-959S.

VAIRAPPAN, C.S. and M. SUZUKI. 2000. Dynamics of total surface bacteria and bacterial species counts during desiccation in the Malaysian sea lettuce, *Ulva reticulata* (Ulvales, Chlorophyta). Phycological Research 48: 55-62.

VAIRAPPAN, C.S, SUZUKI, et.al. 2011. Halogenated metabolites with antibacterial activity from the Okinawan Laurencia species. Phytochemistry 58 (3), 517-523

VALENTE, L.M.P., A. GOUVENIA, P. REMA, J. MALTOS, E.F. GOMES, and I.S. PINTO. 2006. Evaluation of three seaweeds *Gracilaria bursa-pastoris*, *Ulva rigida* and *Gracilaria cornea* as dietary ingredients in European sea bass (*Dicentrarchus labrax*) juveniles. Aquaculture 252: 85-91.

VARGHESE, J.G., A.A. KITTUR, P.S. RACHIPUDI and M.Y. KARIDURAGAN-AVAR. 2010. Synthesis, characterization and pervaporation performance of chitosan-g- polyaniline membranes for the dehydration of isopropanol. Journal of Membrane Science 364 (1-2):10.

VOLESKY, B. and N. KUYUCAK. 1988. Biosorbent for Gold. US. Patent 4, 769,233.

WAALAND, J.R. 1977. Common Seaweeds of the Pacific Coast. J.J. Douglas Ltd., Vancouver, B.C., Canada.

WONG, K.H. and P.C.K. CHEUG. 2000. Nutritional evaluation of some subtropical red and seaweeds Part 1. Proximate composition, amino acid profiles and some physic-chemical properties. Food Chemistry 71: 475-482.

ZENK, M.H. 1996. Heavy metal detoxification in higher plants: a review. Gene 179: 21–30.

ZHOU J.L., P.L. HUANG and R.G. LIN. 1998. Sorption and desorption of Cu and Cd by macroalgae and microalgae. Environmental Pollution 10: 67-75.

http://www.niobioinformatics.In/Seaweeds/System_*Ulva reticulata*.htm

htttp://www.itmonline.org/arts/seaweed.htm

http://www.seaweedireland.com

http://en.wikipedia.org/wiki/Seaweed

http://www.ryandrum.com/seaweeds.htm

http://www.fssai.gov.in